Code of Practice for
Project Management
for Construction and Development

Fourth Edition

The Chartered Institute of Building

CIOB
THE CHARTERED INSTITUTE OF BUILDING

WILEY-BLACKWELL

A John Wiley & Sons, Ltd., Publication

This edition first published 2010
© 2010 The Chartered Institute of Building

Blackwell Publishing was acquired by John Wiley & Sons in February 2007. Blackwell's publishing programme has been merged with Wiley's global Scientific, Technical, and Medical business to form Wiley-Blackwell.

Registered office
John Wiley & Sons Ltd, The Atrium, Southern Gate, Chichester, West Sussex, PO19 8SQ, United Kingdom

Editorial offices
9600 Garsington Road, Oxford, OX4 2DQ, United Kingdom
2121 State Avenue, Ames, Iowa 50014-8300, USA

First published 1992
Second edition 1996
Third edition 2002
Fourth edition 2010

For details of our global editorial offices, for customer services and for information about how to apply for permission to reuse the copyright material in this book please see our website at www.wiley.com/wiley-blackwell.

Library of Congress Cataloging-in-Publication Data
Code of practice for project management for construction and development/the Chartered Institute of Building. — 4th ed.
        p. cm.
    Includes bibliographical references and index.
    ISBN 978-1-4051-9420-4 (pbk. : alk. paper) 1. Building—Superintendence. 2. Project management.
I. Chartered Institute of Building (Great Britain)
    TH438.C626 2010
    690.068—dc22   2009033604

A catalogue record for this book is available from the British Library.

Set in Gothic 720 BT by Gray Publishing, Tunbridge Wells, Kent
Printed and bound in Malaysia by Vivar Printing Sdn Bhd

1   2010

Code of Practice for
Project Management
for Construction and Development

Contents

Part 1 Project management

Part 2 Project handbook

Foreword

It would be fair to say that back in 1992 when the Chartered Institute of Building (CIOB) first produced the *Code of Practice for Project Management* it was a groundbreaking publication.

Whereas modern project management has only been defined over the past couple of decades, the concept of project management has been with us since civilisation began. The need to complete great works and projects within a defined scope, time and cost is not a new idea. It has been central to our past successes, but is just as fundamental for our future achievements.

As an industry that produces many bespoke projects with varying levels of complexity, construction is arguably the world's leading sector when it comes to this kind of management. Our products and the way they are delivered are exemplars for other industries. We lead the way because we challenge ourselves to improve our techniques, innovation and standards constantly.

While this *Code* has been developed specifically for the UK construction industry, its value has been tried and tested in other countries. There has been a steady demand for previous editions from around the world. Indeed the second and third editions have been translated into Chinese and published in China.

This fourth edition of the *Code* continues to drive the practice forward and importantly reflects the current challenges that the industry faces today. As technology has progressed so has its involvement in project management. Making best use of the range of software packages, and e-procurement systems now available is increasingly vital. These products when used appropriately are able to connect even the most complicated networks and processes.

The ingredients which make up modern project management also recognise the need for strong communication that creates valuable relationships with multiple stakeholders. Important agendas like sustainability also have their place in project management. The content within this book provides the reader with helpful and insightful knowledge across a wide range of issues. It is a key reference source for clients, contractors and professionals, irrespective of the size and nature of the project.

To be the best you need ability, experience and knowledge. This essential *Code* provides the latest thinking and guidance for those wanting to achieve that aim.

Professor Li Shirong
President
Chartered Institute of Building

Acknowledgements

I would like to take this opportunity to thank the many people who have helped the Chartered Institute of Building with the fourth edition of this *Code of Practice*. In keeping with the third edition, I am pleased to note that the fourth edition has also been prepared by a broad representation of the industry, with representative contributions from built environment and interdisciplinary co-operation between professionals within the built environment. A list of participants and the organisations represented is included in this book.

Particular thanks must go to Sue Belbin for co-ordinating all the disparate elements of the review of the *Code of Practice* by maintaining the information flow.

This new edition benefited from the capable project management skill and stewardship of Saleem Akram, the CIOB's Director for Construction Innovation and Development, whose efforts to collate and complete this revision deserve a special acknowledgement.

I would also like to thank Arnab Mukherjee for his contributions particularly for technical editing, collating and managing the information flow towards the delivery of this document.

Chris Blythe
Chief Executive
Chartered Institute of Building

Fourth Working Group for the Revision of the *Code of Practice for Project Management*

Saleem Akram BSc Eng (Civil) MSc (CM) PE MASCE MAPM FIE FCIOB — Director, Construction Innovation and Development, CIOB

Alan Crane CBE CEng FICE FCIOB FCMI — Chair, Working Group, Vice President CIOB

Roger Waterhouse MSc FRICS FCIOB FAPM — Vice Chair; Working Group, Royal Institution of Chartered Surveyors Association for Project Management

Neil Powling DipBE FRICS DipProjMan(RICS) — Royal Institution of Chartered Surveyors

Gavin Maxwell-Hart BSc CEng FICE FIHT MCIArb FCIOB — Institution of Civil Engineers

John Campbell BSc (Hons), ARCH DIP AA RIBA — Royal Institute of British Architects

Martyn Best BA Dip Arch RIBA MAPM — Royal Institute of British Architects

Richard Biggs MSc FCIOB MAPM MCMI — Construction Industry Council

Paul Nash MSc MCIOB — Trustee of CIOB

Ian Caldwell BSc BARch RIBA ARIAS MCMI MIOD

Professor James Somerville FCIOB MRICS MAPM MCMI PhD

Professor John Bennett DSc FRICS

David Woolven FCIOB

Artin Hovsepian BSc (Hons) MCIOB MASI

Eric Stokes MCIOB FHEA MRIN

Dr Milan Radosavljevic UDIG PhD MIZS-CEng ICIOB

Arnab Mukherjee BEng(Hons) MSc (CM) MBA MAPM MCIOB — Technical editor

The following also contributed in development of the fourth edition of the *Code of Practice for Project Management*

Keith Pickavance — Past President, CIOB

Howard Prosser CMIOSH MCIOB — Chair, Health & Safety Group, CIOB

Sarah Peace BA (Hons) MSc PhD — Research Manager, CIOB

Mark Russell BSc (Hons) MCIOB — Co-ordinator, Time Management Group, CIOB

Andrzej Minasowicz DSc PhD Eng FCIOB PSMB SIDiR — Vice Director of Construction Affairs, Institute of Construction Engineering and Management, Civil Engineering Faculty, Warsaw University of Technology

John Douglas FIDM FRSA — Englemere Ltd

Dr Paul Sayer — Publisher, Wiley-Blackwell, John Wiley & Sons Ltd, Oxford

List of tables and figures

List of tables

List of figures

Introduction

Project management

Project management has come a long way since its modern introduction to construction projects in the late 1950s. Now, as an established discipline which executively manages the full development process, from the client's idea to funding co-ordination and acquirement of planning and statutory controls approval, sustainability, design delivery, through to the selection and procurement of the project team, construction, commissioning, handover, review, to facilities management co-ordination.

This *Code of Practice* positions the project manager as the client's representative, although the responsibilities may vary from project to project, consequently project management may be defined as 'the overall planning, co-ordination and control of a project from inception to completion aimed at meeting a client's requirements in order to produce a functionally and financially viable project that will be completed safely, on time, within authorised cost and to the required quality standards'.

The fourth edition of this *Code* is the authoritative guide and reference to the principles and practice of project management in construction and development. It will be of value to clients, project management practices and educational establishments and students, and to the construction and development industries. Much of the information contained in the *Code* will also be relevant to project management operating in other commercial spheres.

Raising standards

Project management principles of strategic planning, detailed programming and monitoring, resource allocation and effective risk management are widely used on projects of all sizes and complexity. Nevertheless, many projects do not meet their required performance standards or are delivered late and/or over budget and are sometimes not even 'fit for purpose'. These issues can be addressed directly by raising the standards of project management within the construction and development industries and, more specifically, by improving the skills of project managers in managing the balance of the key constraints of time, quality, cost along with function (end-user requirements) and the important, overriding requirements for sustainability within the built environment (Fig. 0.1).

The three areas of sustainability, economic, social and environmental, have become a core focus for today's projects as the need for greater consideration for the world's climate places an increasing responsibility on all clients and project team members to realise the need for carbon reduction and meaningful, sustainable development.

Past reports such as those by Egan (1998, 2002) and those from cross-industry representative bodies such as the Construction Industry Council (CIC), Construction Industry Research and Information Association (CIRIA), Construction Industry Best Practice Programme (CBPP) and Strategic Forum for Construction (SFC), have collectively made an impact along with similarly focused EU Directives. However, reporting the sustainability message is just the start, informing, training or educating all stakeholders up to the highest level is what is necessary. Effective project management can co-ordinate, broadcast, advise and implement the changes needed, but the raising of this awareness throughout the supply chain and with the client, must come first.

Project managers, as the executive leaders of the project teams, have a responsibility to prepare themselves for this major shift in project thinking, where cost is balanced with the need to conserve energy, minimise emissions, reduce waste, recycle and to construct flexible buildings or facilities with longer life cycles through adaptation and intelligent conversion.

The informed project manager and project teams can encourage the use of modern materials and methods, reduced transportation and greater option appraisals in respect of:

- material choices

- methods of construction

- current account reduction through the use of more efficient equipment

- better insulation materials

- greater recycling of materials.

Continuing professional development or life-long learning is not just about keeping abreast of events; professionals, suppliers, workforce operatives and end users, all need to upgrade their skills and knowledge if we are collectively to achieve the raising of standards needed to meet the new challenges which our industry and environment require.

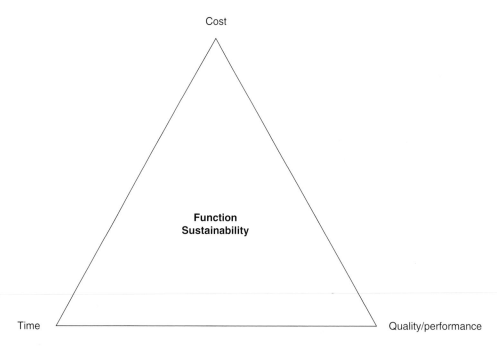

Figure 0.1 Key project constraints.

Adding value

The raising of standards should lead significantly to the adding of value. Greater awareness can result in better design, improved methods and processes, new material choices, less waste, decreases in transportation costs and ultimately more efficient buildings, all of which can bring added value to the whole development process.

The task of project management

Construction and development projects involve the co-ordinated actions of many different professionals and specialists to achieve defined objectives. The task of project management is to bring the professionals and specialists into the project team at the right time to enable them to make their best possible contribution, efficiently.

Professionals and specialists bring knowledge and experience that contributes to decisions, which are embodied in the project information. The different bodies of knowledge and experience all have the potential to make important contributions to decisions at every stage of projects. In construction and development projects there are far too many professionals and specialists involved for it to be practical to bring them all together at every stage. This creates a dilemma because ignoring key bodies of knowledge and experience at any stage may lead to major problems and additional costs for everyone.

The practical way to resolve this dilemma is to carefully structure the way the professionals and specialists bring their knowledge and experience into the project team. The most effective general structure is formed by the eight project stages used in this Code's description of project management.

The different stages of project lifecycle as identified by common terminologies as prevalent across the industry have been summarised and compared in Figure 0.2.

In many projects there will be a body of knowledge and experience in the client organisation which has to be tapped into at the right time and combined with the professional and specialists' expertise.

Each stage in the project process is dominated by the broad body of knowledge and experience that is reflected in the stage name. As described above, essential features of that knowledge and experience need to be taken into account in earlier stages if the best overall outcome should be achieved. The way the professionals and specialists who own that knowledge and experience are brought into the project team at these earlier stages is one issue that needs to be decided during the strategy stage.

The results of each stage influence later stages and it may be necessary to involve the professionals and specialists who undertook earlier stages to explain or review their decisions. Again, the way the professionals and specialists are employed should be decided in principle during the strategy stage.

Each stage relates to specific key decisions. Consequently, many project teams hold a key decision meeting at the end of each stage to confirm that the necessary actions and decisions have been taken and the project can therefore begin the next stage. There is a virtue in producing a consolidated document at the end of each stage that is approved by the client before proceeding to the next stage. This acts as a reference mark as well as acting as a vehicle for widespread ownership of the steps that have been taken.

CIOB Code of Practice for Project Management for Construction and Development	Office of Government Commerce (OGC)	British Standards BS6079-1:2000	British Property Federation (BPF)	Royal Institute of British Architects (RIBA)
1 Inception	Gate 0 Strategic assessment	1 Conception	1 Concept	A Appraisal
2 Feasibility	Gate 1 Business justification	2 Feasibility	2 Preparation of the brief	B Design brief
3 Strategy	Gate 2 Procurement strategy / Gate 3 Investment decision		3 Design development	C Concept
4 Pre-construction		3 Realisation	4 Tender documentation and tendering	D Design development, E Technical design, F Production information, G Tender documentation, H Tender action, J Mobilisation
5 Construction	Gate 4 Readiness for service		5 Construction	K Construction to practical completion
6 Engineering Services Commissioning				
7 Completion, handover and occupation		4 Operation		
8 Post-completion review/project close-out report	Gate 5 Benefits evaluation	5 Termination		L Post-practical completion
	Disposal			

Figure 0.2 Project lifecycle.

Having considered the social, economic and environmental issues, projects begin with the inception stage which starts with the business decisions by the client that suggest a new construction or development project may be required. Essentially, the inception stage consists of commissioning a project manager to undertake the next stage which is to test the feasibility of the project. The feasibility stage is a crucial stage in which all kinds of professionals and specialists may be required to bring many kinds of knowledge and experience into a broad ranging evaluation of feasibility. It establishes the broad objectives, and an approach to sustainability for the project and so exerts an influence throughout subsequent stages.

The next stage is the strategy stage which begins when the project manager is commissioned to lead the project team to undertake the project. This stage requires the project's objectives, an overall strategy and procedures in place to manage the sustainability and environmental issues, and the selection of key team members to be considered in a highly interactive manner. It draws on many different bodies of knowledge and experience and is crucial in determining the success of the project. In addition to selecting an overall strategy and key team members to achieve the project's objectives, it determines the overall procurement approach and sets up the control systems that guide the project through to the final post-completion review and project close-out report stage. In particular the strategy stage establishes the objectives for the control systems. These deal with much more than quality, time and cost. They provide agreed means of controlling value from the client's point of view, monitoring time and financial models that influence the project's success, managing risk, making decisions, holding meetings, maintaining the project's information systems, and all the other control systems necessary for the project to be undertaken efficiently.

At the completion of the strategy stage, everything is in place for the pre-construction stage. This is when the design is developed and the principal decisions are made concerning time, quality and cost management. This stage also includes statutory approvals and consents, considering utility provisions such as water and electricity, monitoring of the environmental performance targets, and bringing manufacturers, contractors and their supply chains into the project team. Like the earlier stages, the pre-construction stage often requires many different professionals and specialists working in creative and highly interactive ways. It is therefore important that this stage is carefully managed using the control systems established during the strategy stage to provide everyone involved with relevant, timely and accurate feedback about their decisions. Completion of this stage provides all the information needed for construction to begin.

The construction stage is when the actual building or other facility that the client needs is produced. In modern practice this is a rapid and efficient assembly process delivering high-quality facilities. It makes considerable demands on the control systems, especially those concerned with time and quality. The complex nature of modern buildings and other facilities and their unique interaction with a specific site means that problems will arise and have to be resolved rapidly. Information systems are tested to the full, design changes have to be managed, construction and fitting out teams have to be brought into the team and empowered to work efficiently. Costs and time have to be controlled within the parameters of project objectives and the product delivered to the quality and specification as set previously.

The construction stage leads seamlessly into a key stage in modern construction and development projects: the engineering services commissioning stage. The complexity and sophistication of modern engineering services makes it essential that time is set aside to test and fine tune each system. Any environmental performance targets such as Building Research Establishment Environmental Assessment Method (BREEAM) certification will be finalised as a measure of the project's performance. Therefore, these activities form a distinct and separate stage which should predominantly be complete before beginning the completion, handover and occupation stage which is when the client takes over the practically completed building or other facility. In some instances there may also be some post-occupation commissioning and testing.

The client's occupational commissioning needs to be managed as carefully as all the other stages because it can have a decisive influence on the project's overall success and environmental performance. New users always have much to learn about what a new building or other facility provides. They need training and help in making best use of their new building or other facility. It is good practice for their interests and concerns to be considered during the earlier stages and preparation for their move into the new facility at the right time so there are no surprises when the client's organisation takes occupation.

The final stage is the post-completion review and project close-out report stage. This provides the opportunity for the project team to consider how well the project's objectives have been met and what lessons should be taken from the project. A formal report describing these matters provides a potentially important contribution to knowledge. For clients who have regular programmes of projects and for project teams that stay together over several projects, such reports provide directly relevant feedback. Even where this is not the case, everyone involved in a project team, including the client, is likely to learn from looking back at their joint performance in a careful objective review.

Part 1
Project management

1

Inception stage

Introduction

Complex capital projects require significant management skills, co-ordination of a wide range of people with different expertise and ensuring completion within the parameters of time, cost and quality specifications. The inception stage of any construction and development project requires the decision from the client that a potential project represents the best way of meeting a defined need.

In assessing the need for construction, key questions should include:

- Why is the project needed?

- How to incorporate sustainability and is the client's corporate responsibility defined?

- How best is the need be fulfilled? (For example, a new building, or refurbishment, or extension of existing structure, etc.)

- What is a reasonable budget cost?

- What is a reasonable time from inception to completion?

- What are the investment and funding options?

- What risks related to the development can be foreseen at this stage?

- What benefits are expected as a result of the project?

Client's objectives

The main objective at this stage for the client is to make the decision to invest in a construction or development project. The client should have prepared a business case (capital expenditure programme) involving careful analysis of its business, organisation, present facilities and future needs. Experienced clients may have the necessary expertise to prepare their business case themselves. Less experienced clients may need help. Many project managers are able to contribute to this process. This process will result in a project-specific statement of need. The client's objective will be to obtain a totally functional facility, which satisfies this need and must not be confused with the project objectives, which will be developed later from the statement of need.

A sound business case prepared at this stage will:

- be driven by needs

- be based on sound information and reasonable estimation

- contain rational processes

- be aware of the risks associated

- contain flexibility

- maximise the scope of obtaining best value from resources

- utilise previous experience

- incorporate sustainability cost-effectively.

Client's internal team

Investment decision-maker: this is typically a corporate team of senior managers and/or directors who review the potential project and monitors the progress. However, the team seldom is involved directly in the project process.

Project sponsor: typically a senior person in the client's organisation, acting as the focal point for key decisions about progress and variations. The project sponsor has to possess the skills to lead and manage the client role, have the authority to take day-to-day decisions and have access to people who are making key decisions.

Client's advisor: the project sponsor can appoint an independent client advisor (also referred to as construction advisor or project advisor or independent client advisor) who will provide professional advice in determining the necessity of construction and means or procurement, if necessary. If advice is taken from a consultant or a contractor, those organisations have a vested interest not only in confirming the client's need, but also in selling their services and products.

The client advisor can assist with:

- business case development (further guidance on this has been provided in Appendix 30)

- investment appraisal

- designing and planning for sustainability

- understanding the need for a project

- deciding the type of project that meets the need

- generating and appraising options (when appropriate)

- selecting an appropriate option (when available)

- risk assessment (when appropriate)

- advising the client on the choice of procurement route

- selecting and appointing the project team

- measuring and monitoring performance (when appropriate).

The client advisor should understand the objectives and requirements of the client but should remain independent and objective in providing advice directly to the client. Other areas where the client may sought independent advice include: chartered accounting, tax and legal aspects, market research, town planning, chartered surveying and investment banking.

Project manager

Project managers can come from a variety of backgrounds, but all will need to have the necessary skills and competencies to manage all aspects of a project from inception to occupation. This role may be fulfilled by a member of the client's organisation or be an external appointment.

Project manager's objectives

The project manager, both acting on behalf of, and representing the client has the duty of *'providing a cost-effective and independent service, selecting, correlating, integrating and managing different disciplines and expertise, to satisfy the objectives and provisions of the project brief from inception to completion. The service provided must be to the client's satisfaction, safeguard his interests at all times, and, where possible, give consideration to the needs of the eventual user of the facility'*.

The key role of the project manager is to motivate, manage, co-ordinate and maintain the morale of the whole project team. This leadership function is essentially about managing people and its importance cannot be overstated. A familiarity with all the other tools and techniques of project management will not compensate for shortcomings in this vital area. Further guidance on the leadership aspect of the project manager's role has been provided in Appendix 21.

In dealing with the project team the project manager has an obligation to recognise and respect the professional codes of the other disciplines and, in particular, the responsibilities of all disciplines to society, the environment and each other. There are differences in the levels of responsibility, authority and job title of the individual responsible for the project, and the terms project manager, project co-ordinator and project administrator are all widely used.

It is essential, in order to ensure an effective and cost-effective service, that the project should be under the direction and control of a competent practitioner with a proven project management track record developed from a construction industry-related professional discipline. This person is designated the project manager and is to be appointed by the client with full responsibility for the project. Having delegated powers at inception, the project manager may exercise, in the closest association with the project team, an executive role throughout the project with appropriate input from the client.

Project manager's duties

The duties of a project manager will vary depending on the client's expertise and requirements, the nature of the project, the timing of the appointment and similar factors. If the client is inexperienced in construction, the project manager may be required to develop his own brief. Whatever the project manager's specific duties in relation to the various stages of a project, there is the continuous duty of exercising control of project time, cost and performance. Such control is achieved through forward thinking and the provision of good information as the basis for decisions for both the project manager and the client. A matrix correlating suggested project management duties and client's requirements is given in Table 1.1.

Table 1.1 Duties of project manager

Duties*	Client's requirements			
	In-house project management		Independent project management	
	Project management	**Project co-ordination**	**Project management**	**Project co-ordination**
Be named in the contract	•		+	
Assist in preparing the project brief	•		•	
Develop project manager's brief	•		•	
Advise on budget/funding/ programme/risk management arrangements	•		+	
Advise on site acquisition, grants and planning	•		+	
Arrange feasibility study and report	•	+	•	+
Develop project strategy	•	+	•	+
Prepare project handbook	•	+	•	+
Develop consultant's briefs	•	+	•	+
Devise project programme	•	+	•	+
Select project team members	•	+	+	+
Establish management structure	•	+	•	+
Co-ordinate design processes	•	+	•	+
Appoint consultants	•		•	+
Arrange insurance and warranties	•	•	•	+
Select procurement system	•	•	•	+
Arrange tender documentation	•	•	•	+
Organise contractor pre-qualification	•	•	•	+
Evaluate tenders	•	•	•	+
Participate in contractor selection	•	•	•	+
Participate in contractor appointment	•	•	•	+
Organise control systems	•	•	•	•
Monitor progress	•	•	•	•
Arrange meetings	•	•	•	•
Authorise payments	•	•	•	+
Organise communication/reporting systems	•	•	•	•
Provide project co-ordination	•	•	•	•
Issue health and safety procedures	•	•	•	•
Address environmental aspects	•	•	•	•
Co-ordinate statutory authorities	•	•	•	•
Monitor budget and changes	•	•	•	•
Develop final account	•	•	•	•
Arrange pre-commissioning/ commissioning	•	•	•	•
Organise handover/occupation	•	•	•	•
Advise on marketing/disposal	•	+	•	+
Organise maintenance manuals	•	•	•	+
Plan for maintenance period	•	•	•	+
Develop maintenance programme/ staff training	•	•	•	+
Plan facilities management	•	•	•	+
Arrange for feedback monitoring	•	•	•	+

*Duties vary by project, and relevant responsibility and authority.

Symbols: (•) = suggested duties; (+) = possible additional duties.

An example of typical terms of engagement for a project manager is given in Appendix 1. It will be subject to modifications to reflect the client's objectives, the nature of the project and contractual requirements.

The term 'project co-ordinator' is applied where the responsibility and authority embrace only part of the project, e.g. pre-construction, construction and handover/migration stages. (For professional indemnity insurance purposes a distinction is made between project management and project co-ordination. When the project manager appoints other consultants the service is defined as project management and when the client appoints other consultants the service is defined as project co-ordination.)

Appointment of project manager

It is advisable to appoint the project manager at the inception stage, so that the project manager can advise and become involved in the option appraisal process. This should ensure professional, competent management co-ordination, monitoring and controlling of the project to its satisfactory completion, in accordance with the client's brief. However, depending on the nature and type of the project and the client's in-house expertise, the project manager could be appointed as late as the start of the strategy stage, but this could deprive him or her of important background information and is therefore not generally recommended.

Project manager and contract administrator

A *contract administrator* (sometimes also referred to as the employer's agent or supervising officer, depending on the specific contract used) may be appointed for the construction and subsequent stages of the project. This post will have a direct contractual responsibility to the client. The roles of project manager and contract administrator are quite different. The role of project manager is comprehensively described in this *Code* and is very clearly defined in the various conditions of engagement and covers all stages from inception to occupation and project close-out. The role, however, may include elements of administration during perhaps all stages of the project; the extent depending on the particular terms of his/her appointment. The function of the contract administrator is a specific appointment in many forms of building contract and therefore relates specifically to the construction, commissioning and completion stages. The function has remained constant for at least 200 years although the title may change to suit the fashion of the times. While the roles of project manager and contract administrator may be kept separate, they may also be (and often are) combined.

Managing people

Project management, although strongly associated with change management and systems, is above all about managing people. It is about the issues of motivating the project team, middle management and the workforce in order to secure compliance with standards of performance and of gaining their commitment to success. It is also about achieving an effective form of relationship, which will facilitate an atmosphere of mutual trust and co-operation.

People: the most important resource

Although it is important to exploit new technology, computers and software can only do what they are told to do, it is the human resource that will make the difference and ultimately create the competitive advantage. Even computer-based systems are only as good as their designers and operators. People are the industry's most important resource. It requires special skills to be successful at organising, motivating and negotiating with people. Although some people have a greater natural talent for this than others, everyone can improve their natural ability through appropriate education and training.

The skills the project manager will need to consider when assessing an individual may include the following:

- What a person can do: skills, competencies.

- What a person can achieve: output, performance.

- How a person behaves: personality, attitudes, intellect.

- What a person knows: knowledge, experience.

The skills the project manager will use during the course of a project will include:

- Communication: using all means, the foremost skill.

- Organising: using systems and good management techniques.

- Planning: via accurate forecasting and programming.

- Co-ordination: by liaising, harmonising and understanding.

- Controlling: via monitoring and response techniques.

- Leadership: by example.

- Delegation: through trust.

- Negotiation: by reason.

- Motivation: through appropriate incentives.

- Initiative: by performance.

- Judgement: through experience and intellect.

Establishing objectives

The recognition that members of the project team have differing and sometimes conflicting objectives is the first step in ensuring that the team operates as an effective unit. With the client's project strategy in sharp focus, attention is directed towards overcoming any conflict in the aims of team members. Presentation of objectives, team selection, choice of working environment, definition of levels of responsibility, authority and communication procedures; are all important in ensuring that team members meet their personal ambitions as part of the successful execution of the project.

The project manager should aim to create an environment in which the client and all his team members can achieve their personal, as well as project, goals. There is no doubt that team performance is optimised when members are encouraged to identify and tackle problems early in the process. Promotion of an open, 'blame-free' culture, where the project manager leads by example, will also help in breaking down communication barriers.

Thinking sustainably

The sustainability of the development and the client's corporate responsibility have already been touched on earlier in this chapter. It is important at this early stage of the development process that the message concerning the triple bottom line – environment (planet), economy (profit) and society (people) – is considered seriously.

Projects need to be financed and designed from the very beginning with sustainability high on the agenda, for it is at this stage when plans can be formulated holistically, and where the greatest cost benefits can be derived. It is here where ideas are made to not just achieve minimum environmental and social standards, but to advance them so that maximum capital and whole-life costs can be achieved.

Although the project has not yet reached feasibility, ideas should be formulating for their introduction at the feasibility stage:

■ What kind of energy sources?

■ What kind of mechanical and electrical systems and which manufacturers are achieving the highest efficiencies?

■ How much can be obtained locally to minimise transport (carbon) consumption?

■ What are the standards not only likely to be needed now, but in the foreseeable future?

This may be just the 'idea' stage of a project, but thinking sustainably now is the way to build in lower costs and increase value.

All participants will eventually need to be informed and invited to participate in helping to achieve the project's aims, procedures and targets for reducing emissions as soon as they enter the site. Information leaflets and induction classes will need to be made available at the same time as the presentations are made for health and safety.

Measures identifying how energy is to be conserved and recorded during each activity on and off site can be identified and considered, so that carbon and cost consumptions can be monitored and the effects of new methods of conservation determined. These processes for monitoring all energy consumed, both fossil sourced and renewable, directly and indirectly, can be incorporated into a site energy plan and carbon register so that the minimum emissions from the project can be achieved. Concepts such as these can be considered here, although details will not be finalised until a later stage.

2 Feasibility stage

Client's objectives

The main objectives for the client at this stage include identifying and specifying the project objectives, outlining possible options and select the most suitable option through sustainability, value and risk assessment. At this stage establishing the project execution plan (PEP) for the selected option should be the key output.

Outline project brief

For most clients a building is not an end in itself, but merely the means to an end: the client's objectives. The client's objectives may be as complex as the introduction and accommodation of some new technology into a manufacturing facility or the creation of a new corporate headquarters; or they may be as simple as obtaining the optimum return on resources available for investment in a speculative office building.

The client's objectives are usually formulated by the organisation's board or policy-making body (the investment decision-maker) and may include certain constraints – usually related to time, cost, performance and location. The client's objectives must cover the function and quality of the building or other facility.

If it is considered that the objectives are complex enough to merit the engagement of a project manager, the appointment should ideally be made as early as possible, preferably after approving the project requirements at the inception stage. This will ensure the benefit of the special expertise of the project manager in helping to define the parameters and in devising and assessing options for the achievement of the objectives.

The project manager should be provided with, or assist in, preparing a clear statement of the client's objectives and any known constraints. This is the initial outline project brief to which the project manager will then work. A typical example of a template for an outline project brief is shown in Figure 2.1.

PROJECT TITLE

PROJECT REF

CUSTOMER: *(internal/external)*

PROJECT SPONSOR:

PROJECT MANAGER:

GOAL

THIS NEEDS TO BE SPECIFIC AND INCLUDE THE JUSTIFICATION FOR THE PROJECT

It should spell out: **what** *will be done and by* **when**;

OBJECTIVES

It is essential these cover the OUTCOMES expected of the project and that preferably they are:

Specific – i.e. clear and relevant

Measurable – i.e. so it is feasible to see when it is happening

Achievable/agreed to – helpful to use positive language and that others 'buy-in' to the objectives

Realistic – this depends on three factors: resources/time/outcome or aim

Time bound – have a time limit: without this they are wishes

APPROACH

The project plan should include the key milestones for the review, i.e. set a target date for agreeing the project brief and target dates for completing key stages of the project.

SCOPE

THIS SETS THE PROJECT BOUNDARIES AND IT CAN BE USEFUL TO ADD WHAT IS NOT COVERED.

It can be a useful reference point if the project changes in due course

CONSTRAINTS

Could add 'start' date and 'end' dates here

It is particularly important in the context of best value, to identify here genuine constraints rather than customer preconceived ideas about the solution

DEPENDENCIES

This identifies factors outside the control of the project manager, and may include:

- Supply of information
- Decisions being taken at the right time
- Other supporting projects

RESOURCE REQUIREMENTS

Include estimate of project days and costs

AGREED

Signature: Date:

Project manager:

Project sponsor:

Note: The above example of the possible template of an outline project brief is for guidance purposes only.

Figure 2.1 Outline project brief.

Part 1 Project management

Feasibility studies

There is seldom, if ever, a single route available for the achievement of the client's objectives, so the project manager's task is to work under the client's direction to help establish a route which will best meet the client's objectives within the perceived constraints that are set. In liaison with the client, the project manager will discuss the available options and initiate feasibility studies to determine the strategy to be adopted. In order that the feasibility studies are effective, the information used should be as full and accurate as possible. Much of that information will need to be provided by specialists and experts. Some of these experts may be available within the client's own organisation or be regularly retained by the client: lawyers, financial advisors, insurance consultants and the like. Others, such as architects, engineers, quantity surveyors, project planners, planning supervisors, town planning consultants, land surveyors and geotechnical engineers may need to be specially commissioned. In some instances it is desirable to involve constructors (e.g. in case of framework contracts or even design and build contracts) in preparation and completion of the feasibility study.

Feasibility study reports should include the following:

■ Scope of investigation (from outline project brief) including establishing service objectives and financial objectives.

■ Studies on requirements and risks.

■ Public consultation (if applicable).

■ Consultation with stakeholders and third parties.

■ A geotechnical study (if applicable).

■ Environmental performance targets (e.g. BREEAM, Code for Sustainable Homes and EcoHomes; refer to Appendices 31 and 32).

■ An environmental impact assessment (refer to Appendix 11).

■ A health and safety study.

■ Legal/statutory/planning requirements or constraints.

■ Risk management strategy.

■ Estimates of capitals and operating costs (demolition costs, if applicable).

■ Assessment of potential funding.

■ Potential site assessments (if applicable).

■ Outline schedule.

The client will commission feasibility studies and establish that the project is both financially viable and deliverable. The client may instruct the project manager at this stage and, if so, his input will be made alongside the reports and views of the various consultants.

The client may ask the project manager to engage and brief the various specialists for the feasibility studies, co-ordinate the information, assess the various options and report on his conclusions and recommendations. The feasibility report should include as a minimum a 'risk assessment' for each option and will usually also determine the contractual procurement route to be adopted and an outline development schedule applicable to each. The client may also require comparative 'lifecycle costings' to be included for each option. The issue of sustainability is now a significant part of all development projects, whether new build or refurbishment, and the three elements of environment, economy and society must be considered from the earliest stages of each project. Further guidance to prepare the feasibility report as part of the business case is available in Appendix 30.

During the progress of the feasibility studies, the project manager will convene and minute meetings of the feasibility team, report progress to the client and advise the client if the agreed budget is likely to be exceeded. Feasibility studies including revenue assessment are the most crucial, but also the least certain phase of a project. Time and money spent at this stage will be repaid in the overall success of the project. The specialists engaged for the feasibility studies are most commonly reimbursed on a time–charge basis and without commitment to engage the specialist beyond the completion of the feasibility study, although often some or all members of the feasibility team will be invited to participate in the selection process to become design team members.

The project manager will obtain a decision from the client on which option to adopt for the project and this option is designated the outline project brief. The process of developing the project brief from the client's objectives is shown in Figure 2.2.

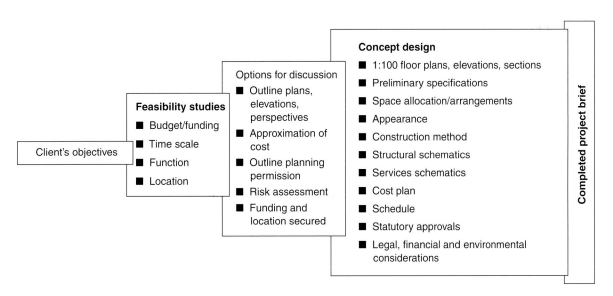

Figure 2.2 Development of project brief from objectives.

Sustainability in the built environment

The impact on the environment of construction and related activities is now well publicised and governments around the world are challenging construction processes, methods, practices and procedures and are monitoring the industry's outputs. The aim is to both question and change the way modern buildings are constructed and used. It is expected that such changes to processes and so forth will have a great impact on both embodied energy and also on energy input requirements of buildings, both existing and new and thus positively impact global emissions of greenhouse gases. Such change will not come cheap as construction processes are adapted and building standards and regulations set new challenges to meet future human requirements and governmental goals.

Towards sustainable development

There is a tendency by many people to consider 'sustainability' as just the 'environment', perhaps because there is so much written about climate change and the emphasis on reducing carbon emissions and pollution. It is these drivers that appear to be leading the sustainability agenda. However, the other two important elements of the 'triple bottom line' which comprise 'sustainability', are social (people) and economic (profit). Of these it is the latter – economic – which many believe is the greater concern, for saving the environment is not expected to be cheap and it is only natural that most stakeholders want to protect their profits.

The Stern report believes extreme weather, created by global warming, could reduce global GDP by up to 1% with the worse-case scenario predicting global output per head falling by 20%. This cost of 'doing nothing' however can arguably be much greater than the costs associated with reducing carbon emissions, although there are many who show reluctance to the implementation of the preventative actions needed to reduce such emissions, on the grounds of associated costs and hence loss of profit.

However, many consultants are successfully demonstrating that if sustainability is approached holistically, using integrated design with modern materials and systems, it is entirely possible to create a built environment that is both sustainable and economically viable. Furthermore, the argument now held by the majority of industry experts is that we really have no choice, the planet is in need of environmental adjustment and the only way for this to happen is by reducing carbon emissions through changes to our built environment, our energy sources and consumptions, transport systems and lifestyles. But such changes are not simple in today's world of politics and limited resources, consequently there is a need to consider holistically the 'triple bottom line' of environment, economic and social, which means that profit and people must remain as key considerations of the overall solution.

In terms of social (people) acceptance, the construction industry needs to build facilities which do not adversely affect the external environment through pollution, excessive size, unsightly design, overcrowding, high maintenance and indulgent consumption of materials and resources. At the same time, homes must be affordable and available, with adequate communal green areas and spaces within environments which are comfortable in terms of light, noise, density and within reasonable commuting distances of workplaces. Travel is a key source of carbon pollution and natural resource consumption, hence creating shorter commuting distances is part of the overall aims for developing a healthier environment.

Offices and shops should follow similar guidelines by being socially acceptable and within a built environment which encompasses safety, ease of access to transport and entertainment facilities, with well-designed structures and community engagement facilities designed to create healthy lifestyles and spiritual well-being.

With regard to the economic (profit) element, making buildings and facilities financial viable while being environmentally friendly is the ultimate goal for the construction industry. Unless there is careful co-ordination of the design, manufacture, choice of materials and specification, construction and operational efficiency, it can be very difficult to create environmentally and socially acceptable buildings within realistic budget limits. This is particularly true when refurbishments are undertaken.

Returning to the environment, as indicated earlier, saving the planet is not an option if we wish to maintain the current balance of nature, we must abandon wasteful practices and produce facilities which are sustainable, socially acceptable and which protect our natural resources without destroying the environment. However, it is important to create comfortable internal environments which are constructed and operated efficiently, while producing financial returns which are also sustainable. To achieve such harmonious environments it is important to follow the national and internationally agreed standards, i.e. BREEAM, LEED, Green-Star, etc. This will help reduce consumer demand for heavily polluting goods and services, the aim being to promote cleaner energy and transport systems with non-fossil fuels producing at least 60% of the required energy output by 2050 in order to achieve the required drastic reductions of carbon emissions.

Potential for government carbon controls

It is anticipated that eventually governments are likely to insist on carbon footprint measurements for each project period, bearing in mind that major projects can take many years to complete. The selection of materials such as concrete and steel, which have high carbon manufacturing consumption as well as potentially high transportation costs, might well bring about the use of alternative materials. Systems such as mechanical and electrical equipment and cladding may well be carbon assessed and their location of manufacture, their indirect and embodied carbon including transportation and the operational carbon consumptions, are likely to be similarly assessed and weighted against initial capital costs, when determining the choice of supplier.

Responsible sustainable development has therefore become a growing national and international issue and project managers and clients should co-operate to make sure that the whole supply chain is aware of its duties during the early stages of each project. The project manager in particular, needs to assure the client that proper consideration by the design and construction teams will be given to the sustainability aspects of the project and its construction process, so that the final building will have a minimal detrimental, effect on the natural environment. To help achieve this, the project team will need to adopt innovative methods and best practice. Therefore, controlling the construction process by effective use of advanced management tools and systems (see below) should consider the carbon footprint of the whole construction process from inception to completion, although efficient monitoring of such tools will need to be undertaken along with the measurement of the associated impact being experienced.

Typical management tools and systems include the following:

■ Detailed site waste management plans (SWMPs) with emphasis on minimum waste generation on all projects, in particular those over £300,000.

■ Use of competent ICT with a top-down approach to managing sub-contractors via the use of inter-operable software systems.

■ Use of electronic document management systems with secure databases.

■ Incorporation of building information modelling (BIM) system approaches to design from an early stage in the process.

■ The use of e-tendering and e-commerce solutions approaches to project management in the pre-construction and construction phases.

- Installation of effective wireless technologies and RFID data collection devices.

- Instant messaging and the use of virtual office(s) and conferencing facilities as a means of reducing the need for travel between project team members.

- Ensuring accurate whole-life costs (WLC) are applied to the project building so that clients can make the best judgements at all stages.

Achieving sustainable development

It is anticipated that eventually regulators are likely to insist on carbon footprint measurements for the period of projects, bearing in mind that major projects can take many years to complete. The selection of materials such as concrete and steel, which have high carbon manufacturing costs as well as potentially high transportation costs might well see the use of alternative materials. Systems such as mechanical and electrical equipment and cladding may well be carbon assessed and the location of manufacture with its associated direct, indirect and embodied carbon consumption including transportation as well as the operational carbon consumption might well be weighed against initial capital costs.

Sustainable development in terms of design and construction imposes duties on all those involved on the entire project at all stages from inception to completion and then through the life of the development to the grave (and then back again).

Appendix 31 provides further information on key sustainability issues and actions at each project stage and Appendix 32 provides further information on BREEAM, Code for Sustainable Homes and EcoHomes.

Most clients and government departments are proactively engaged and committed to securing sustainable development. They require the construction industry to respond to the greater demand for social, economic and environmental improvements such as living within the planet's environmental resources and creating biodiversity, ensuring a strong, healthy and just society with equal opportunities for all, building a strong and stable economy, using sound science for advancement and promoting good governance.

Sustainable development (or sustainability) is defined in BS8900 as: 'an enduring, balanced approach to economic activity, environmental responsibility and social progress'. In CIBSE's introduction to sustainability it is about enabling: 'all people throughout the world to satisfy their basic needs and enjoy a better quality of life without compromising the quality of life for future generations'. In the Report of the World Commission on Environment and Development, *Our Common Future* (The Brundtland Report), it is the development 'that meets the needs of the present without compromising the ability of future generations to meet their own needs'.

Sustainable development is frequently defined as the interaction of social, economic and environmental (ecological) issues. This is represented in Figure 2.3.

Figure 2.3 A summary of sustainable development. Adapted from CIRIA C571 'Sustainable construction procurement: a guide to delivering environmentally responsible projects.

Site selection and acquisition

Site selection and acquisition is an important stage in the project cycle where the client does not own the site to be developed. It should be effected as early as possible and, ideally, in parallel with the feasibility study. (It is to be noted that the credibility of the feasibility study will depend on the major site characteristics.) The work is carried out by a specialist consultant and lawyers and may involve a substantial due diligence exercise. This will be monitored by the project manager.

The objectives are to ensure that the requirements for the site are defined in terms of the facility to be constructed, that the selected site meets these requirements and that it is acquired within the constraints of the outline development schedule and with minimal risk to the client.

To achieve these objectives the following tasks will need to be carried out:

■ Preparing a statement of objectives/requirements for the site and facility/buildings and agreeing this with the client.

■ Preparing a specification for site selection and criteria for evaluating sites based on the objectives/requirements.

■ Establishing the outline funding arrangements.

■ Determining responsibilities within the project team (client/project manager/commercial estate agent).

■ Appointing/briefing members of the team and developing a schedule for site selection and acquisition; monitoring and controlling progress against it.

■ Actioning site searches and collecting data on sites, including local planning requirements, for evaluation against established criteria.

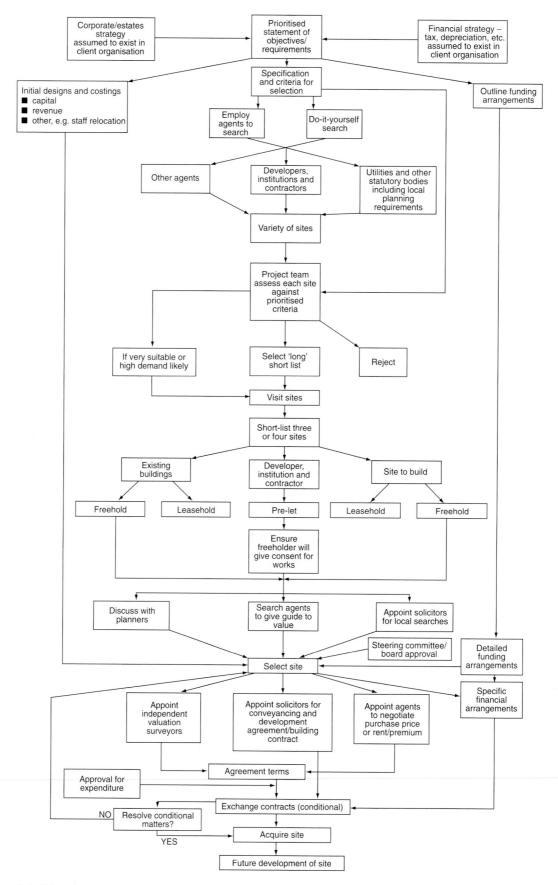

Figure 2.4 Site selection and acquisition.

- Evaluating sites against criteria and producing a shortlist of three or four; agreeing weightings with the client.

- Establishing initial outline designs and developing costs.

- Discussing short-listed sites with relevant planning authorities.

- Obtaining advice on approximate open-market value of short-listed sites.

- Selecting the site from a shortlist.

- Appointing agents for price negotiation and separate agents for independent valuation.

- Appointing solicitors as appropriate.

- Determining specific financial arrangements.

- Exchanging contracts for site acquisition once terms are agreed, conditional on relevant matters, e.g. ground investigation, planning consent.

Project brief

The formulation of the brief for the project is an interactive process involving most members of the design team and appropriate representatives of the client organisation. It is for the project manager to manage the process, resolving conflicts, obtaining client's decisions, recording the brief and obtaining the client's approval. Managing the client organisation to ensure input into the brief comes from the right person is as important as it is time consuming. This ensures that the project manager gets to know and understand the structure, culture and personalities of the client body. At best this establishes ownership and champions for various aspects of the project at an early stage. Table 2.1 lists some suggested contents for a project brief.

Table 2.1 Contents for project brief

The following is a suggested list of contents, which should be tailored to the requirements and environment of each project.

- Background
- Project definition, explaining what the project needs to achieve. It will contain:
    1. Project objectives
    2. Project scope
    3. Outline project deliverables and/or desired outcomes
    4. Any exclusions
    5. Constraints
    6. Interfaces
- Outline business case
    1. A description of how this project supports business strategy, plans or programmes
    2. The reason for selection of this solution
- Customer's quality expectations
- Acceptance criteria
- Risk assessment

If earlier work has been done, the project brief may refer to the document(s) containing useful information, such as the outline project brief, rather than include copies of them. It is not unusual during this phase for the client to modify his thinking on various aspects of the proposals, and there is certainly the opportunity and scope for change during this phase. Figure 2.5 demonstrates graphically the relationship between 'scope for change' and the 'cost of change' set against the time-scale of a development. It will be seen that the crossover point occurs at the completion of the strategy stage. The Client's attention should always be drawn to this relationship and to the benefits of brief and design freezes.

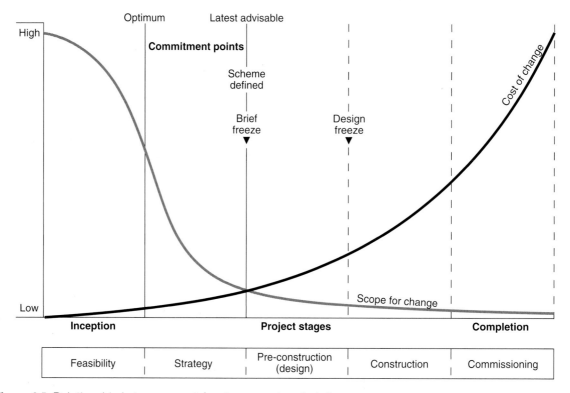

Figure 2.5 Relationship between scope for change and cost of change.

The key emphasis for the client should be to understand and establish enough information about the end requirements and objectives for developing the project. This point cannot be overemphasised. It is essential that the project manager identifies the client's needs and objectives through careful and tactful examination, in order to minimise the risk of potential future changes to the project brief. Many clients who are unfamiliar with the development process are not perhaps fully aware of the importance of getting the design right as much as feasible at the start of the project, and therefore the importance of making sure the brief fully reflects the client's requirements, before design commences and getting the design right before materials ordering and construction commences. The project manager should do his utmost, therefore, to familiarise the client with the potential cost and time implications of design changes and identify as clearly as possible the precise requirements of the client. While this input may come from the project manager, it is advisable, especially for complex and business critical projects, that a 'client advisor' is appointed to give independent advice until the need for a construction project has been established.

Design brief

Within the project brief, the assembly of the design brief will normally be the responsibility of the lead design consultant along with the project manager and, where appropriate, the client and the constructors. The project manager will monitor the assembly of the design brief to ensure compliance with the outline project brief, the project budget and the master programme.

Depending on the nature of the project, procurement method adopted and the master programme, it is sometimes practical for some activities to proceed in parallel. However, if the design brief is not finalised before the final concept design is started, then change, delay and increased costs are almost bound to follow. The site preparation should never start until after the concept design has been finalised and signed off. While decisions on some elements of the design brief may be deferred by the client even until after construction has commenced, this inevitably involves some risk for which time and cost contingencies should be provided. Accordingly, in order to promote certainty, it is much better and usually possible (except in extreme circumstances) to complete the brief and the design before construction begins. It just needs good project management.

The project manager will advise the client of the implications for cost, time and risk in the deferment of any elements of the design brief. The project manager will monitor the progress of the assembly of the concept design and notify the client of the effects on cost, time, quality, function and financial viability of any changes from the design brief.

Funding and investment appraisal

In all development projects, a balance between cost and value must be established. The financial appraisal of the project can either be assessed by calculating the total cost and then assessing the value or alternatively, calculating the value of the end product and working out the project costs with an eye to value. In either case, the client will expect value to exceed cost, and in the case of developer-led projects the client will at inception stage have decided on the level of profit (or benefit) required for the amount of risk involved. A thorough risk analysis, particularly analysing the market conditions on the potential revenue generation, interest rate changes, potential impact of programme delay and outcomes of similar historical precedents (which may be incorporated within the business plan or development appraisal for the project) is usually performed to assist in decision-making. Developers and many clients experienced in construction procurement may not require specific help from the project manager in these areas but should keep him well informed of the financial arrangements so that they can be taken into account in any project decisions. On the other hand, clients unfamiliar with construction may require an input by the project manager or from another independent advisor. In any case, although the project manager may have knowledge of project finance, it is unlikely that he will be expected to advise in this area. Specialist advisors or the client himself will arrange bank finance; take tax and legal advice in all those areas relating to the acquisition of the site and the financing of the development project. The project manager should be able to advise on certain matters relating to VAT, budgetary systems, cost and cash flow. The project manager should also know when and where to go for specialist advice to augment his own expertise or his client's expertise in such matters.

Market suitability

The key to a successful project is to try to bring together all the various elements into a workable and viable whole. Market awareness and being able to judge not only occupiers' requirements but also trends in the investment market are absolutely vital. A key issue at this stage is to ensure appropriate site selection keeping in mind the market that is being addressed.

Decision to go ahead

The client, reviewing the documents generated throughout this phase, has to reaffirm the decision to proceed with the project, in order to:

■ Provide the authorisation of financial management and control throughout the project.

■ Ensure that no commitment is made to large expenditure on the project before verifying that it is sensible to do so.

Table 2.2 Client's decision prompt list

> ■ Is there adequate funding for the project?
> ■ Does the detail project brief demonstrate the existence of a worthwhile project, and hence justify the investment involved?
> ■ Are external support and facility requirements available?
> ■ Have the most appropriate standards been applied, in order to achieve the best value for money?
> ■ Are assurance responsibilities allocated and accepted?
> ■ Is the development sustainable?

At the early stages of a project it is unlikely that the actual costs will be known. It is important to check the need for financial provision (including time, cost and risk contingencies) has been recognised by all the parties. The predicted project cost at later stages, which will in most cases be different to the original estimate, requires the question of affordability to be revisited at that stage to be sure that adequate funds are available.

Project execution plan

The project execution plan (PEP) is the core document for the management of a project. It is a statement of policies and procedures defined by the project sponsor; although it is usually developed by the project manager for the project sponsor's approval. It sets out in a structured format the project scope, objectives and relative priorities.

This is a live document that enforces discipline and planning having a wider circulation than the project design team. It forms a basis for:

■ Sign off by the client body at the end of the feasibility and strategy phases.

■ An aid for funding.

■ A *modus operandum* for the project team.

■ An information and 'catch up' document for prospective contractors.

Some of the confidential information in the client version will be taken out of the published version to other parties.

Checklist for the PEP

Does the PEP:

■ Include plans, procedures and control processes for project implementation and for monitoring and reporting progress?

- Define the role and responsibilities of all project participants, and is a means of ensuring that everyone understands, accepts and carries out their responsibilities?

- Set out the mechanisms for quality control, audit, review and feedback, by defining the reporting and meeting requirements, and, where appropriate, the criteria for independent external review?

Essential contents

Much of a PEP will be standardised, but the standard will need to be modified to meet the particular circumstances of each project. A typical PEP might cover the items listed below, although some may appear under a number of headings with a cross-reference system employed to avoid duplication:

- Project definition and brief.

- Statement of objective.

- The business plan with costs, revenues and cash flow projections including borrowings interest and tax calculations.

- Market predictions and assumptions in respect of revenue and returns.

- Functional and aesthetic brief.

- Client management and limits of authority including the project manager.

- Financial procedures and delegated authority to place orders.

- Development strategy and procurement route.

- Statutory approvals.

- Risk assessment.

- Project planning and phasing.

- The scope content of each consultant appointment.

- Reconciled concept design and budget.

- Method statement for design development, package design and tendering, construction, commissioning and handover, and operation.

- Safety and environmental issues, such as the construction design and management regulations, carbon dioxide emissions and energy targets.

- Management of information systems including document management systems.

- Quality assurance.

- Post-project evaluation.

The PEP will change as a project progresses through its design and construction stages. It should be a dynamic document regularly updated and referred to as a communication tool as well as a control reference.

3 Strategy stage

Client's objectives

The main aims for the client at this stage include setting up the project organisation, establishing the strategies for procurement, delivery (cost, time and quality control and risk management) and commissioning/occupation issues through identifying project targets, assessing and managing risks and establishing the project plan.

Interlinking with feasibility

Distinction between the tasks and activities of the feasibility and strategy stages is not always clear, as each is influenced to a certain extent by the considerations and findings of the other. The two tasks and activities need to relate to each other in order to achieve effective outcomes for both. Feedback (Figure 3.1) is essential in order to establish for the client a sound basis for decision-making at the conceptual phases of the project and, subsequently, for its effective execution. The order in which the activities are set out here is not significant and will vary for specific projects.

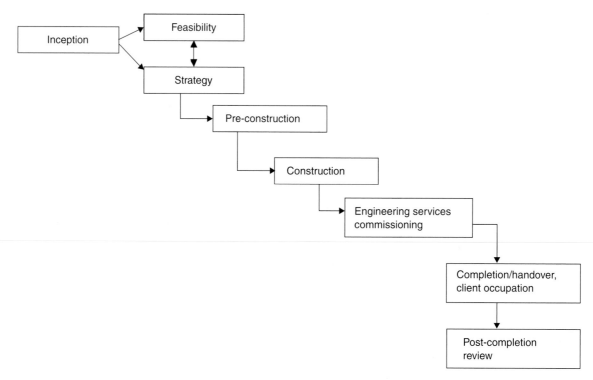

Figure 3.1 Stages of the project development.

Project team structure

Under the overall direction and supervision of a project manager, projects are usually carried out by a project team. The team normally comprises of the following:

■ client's internal team (appropriate representatives)

■ project manager (either within the client's own organisation or independently appointed)

■ design team: architects, structural/civil/mechanical and electrical (M&E) engineers and technology specialists

■ consultants covering quantity surveying, development surveying, planning and scheduling, legal issues, valuation, finance/leasing, insurances, design audit, sustainability and energy certification, health and safety and environmental protection, access issues, facilities management, highways/traffic planning, construction management, and other specialisms

■ contractors and subcontractors.

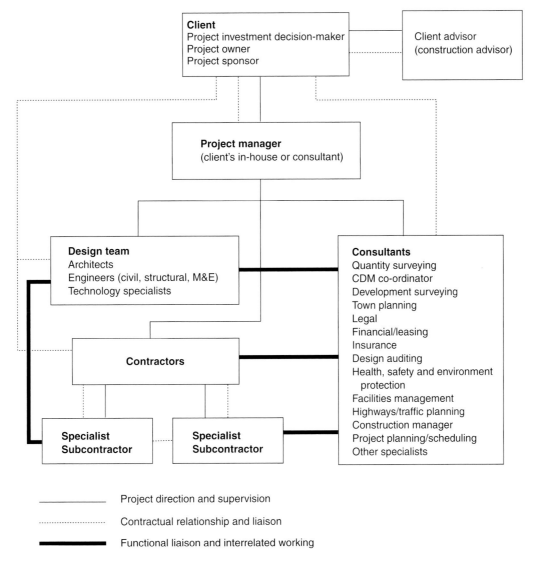

Figure 3.2 Project team structure.

The project team structure for project management is shown in Figure 3.2. This structure is idealised and, in practice, there will be many variants depending on the nature of the project, the contractual arrangements, type of project management (external or in-house) involved, and above all, the client's requirements. It should be one of the duties of the project manager to advise the client on the most appropriate project team structure for a particular project.

Effective project management must, at all times, fully embrace all provisions for quality assurance, time and financial control, health and safety, access provision and environmental protection. These aspects are to be considered as incorporated and implied in all relevant activities specified in this *Code of Practice*.

Selecting the project team

When establishing a project team, many skills will be needed. During selection the project manager should consider the following factors:

■ A commitment by the project team to clearly defined and measurable project objectives.

■ Firm duties of teamwork, with shared financial motivation to pursue those objectives. These should involve a general presumption to achieve 'win-win' solutions to problems which may arise during the course of the project. Issues such as leadership, communication and teamworking form key cornerstones of a successful project delivery. Further guidance on these issues has been detailed in Appendix 21.

■ The production of satisfactory evidence from each team member, to show that they can contribute effectively to the project objectives. This evidence may include a realistic schedule with appropriate allocation of contingencies against foreseeable risks, a financial plan and a demonstration of adequate resources.

■ When choosing each team member, as suggested in Chapter 1, special attention to be paid to their:

    ○ relevant experience

    ○ technical qualifications

    ○ appreciation of project objectives

    ○ level of available supporting resources

    ○ creative/innovative ability

    ○ enthusiasm and commitment

    ○ positive team attitude

    ○ communication skills.

■ Financial strength and core resource strength are also important.

■ Defining clear lines of communication between the respective project team members.

■ Promoting a working environment that encourages an interchange of ideas by rewarding initiatives which ultimately benefit the project.

■ Undertaking regular performance appraisals for all project team members.

■ Ensuring that project team members are suitably located and that communication protocols have been established (particularly for electronic sharing of information) so as to facilitate regular contact with each other, as well as with their own organisations.

■ Defining clear areas of responsibility and lines of authority for each project team member, and communicating these within the team.

■ Identifying a suitable deputy for each team member, who will be sufficiently familiar with the project to be able to act as their replacement should the need arise.

■ Making provision for members of the project team to meet informally and socially, outside the work environment, on a regular basis.

Strategy outline and development

A typical strategy stage consists of the main elements shown in Figure 3.3.

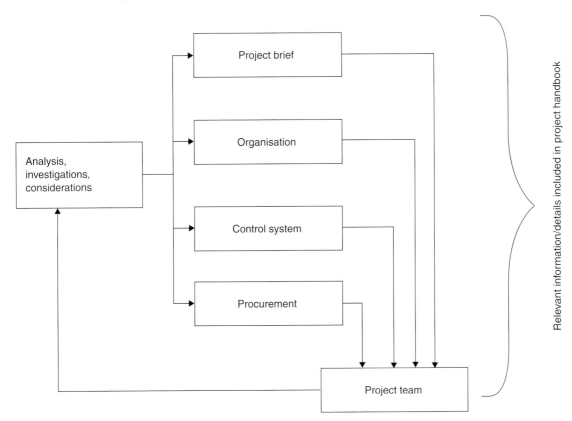

Figure 3.3 Elements of the strategy stage.

The project manager performs several principal activities at this stage which may include all, or most of the following:

■ Reviewing, and in some cases developing, the detail project brief with the client and any existing members of the project team to ascertain that the client's objectives will be met. Preparing a final version in written form with supplementary appendices where these add to the general understanding of the issues that support the brief itself.

■ Establishing, in consultation with the client and consultants, a project management structure (organisation) and the participants' roles and responsibilities, including access to client and related communication routes, and 'decision required' points. This should be developed and presented in the project documents for the reference of all parties.

■ Ensuring, in liaison with the client, the CDM co-ordinator, design consultants and the principal contractor when appointed, that appropriate arrangements have been made to meet the requirements of the Construction (Design and Management) (CDM) Regulations. Key duties under the regulations are summarised in Appendix 2.

■ Developing the client's performance brief for environmental sustainability including targets for recognised assessments such as BREEAM, LEED and Energy Performance, determining how this will be assessed and whether a specialist consultant such as a BREEAM assessor should be appointed. As part of this, establishing investment criteria for environmental measures and including them in value management reviews.

■ Establishing that 'risk and value management' principles are applied effectively from the earliest stages of the preparation of the design brief until the design is complete. The emphasis should be on providing value for money and in producing a facility that can be constructed and operated at the optimum time and cost without compromising quality, scope or specifications. The design team and consultants should be encouraged not to accept conventional wisdom on how long buildings take to build or what buildings/ facilities cost, but consciously seek to identify the optimum balance of time and cost by better design and construction logistics. An approach where the whole team 'designs in quality and time-effectiveness and drives out cost' at all stages in the design process should be encouraged – emphasis on overall value should be encouraged. Further guidance on value management is included Chapter 4 and in Appendix 10 (also see the CIOB's *Guide to Risk and Value Management* and *Managing Time in Complex Projects*).

■ Advising the client on the recruitment and appointment of additional consultants and design team members, i.e.

    ○   preparation of appropriate definition of roles and responsibilities

    ○   preparation and issue of selection/tender documentation

    ○   evaluation, reporting and making recommendations

    ○   assisting the client in the preparation of agreements and in selection and appointment.

■ Drawing the client's attention to the benefits of project insurance for the whole team and works and providing assistance in assessing the risks on the project and including an appropriate contingency sum in the project budget and time contingencies in the outline development schedule. Putting in place procedures for managing risk as a continuous project activity. A project risk assessment checklist (Appendix 9) may be used or adapted as part of such a procedure. (These risks are not to be confused with risks covered by the CDM Regulations, although CDM risks will form a subset of an overall risk management regime.)

■ Selection, or development, and agreement of the most appropriate form of contract relative to the project objectives and the parameters of cost, time, quality, function and financial viability and risk management.

■ Assisting the client in completing site selection/evaluation, investigation and acquisition.

■ Advising on whether certain activities, such as fitting out and occupation/ migration, constitute separate projects and should be treated as such.

■ Making the client aware of relevant statutory submissions and other consultations that may be required in the delivery of the project.

Project organisation and control

A project management organisation structure sets out unambiguously and in detail how the parties to the project are to perform their functions in relation to each other in contributing to the overall scheme. This should be recorded in the *Project Handbook* (see Glossary for definition). It also identifies arrangements and procedures for monitoring and controlling the relevant administrative details. It is updated as circumstances dictate during the lifetime of the project, and should allow project objectives and success criteria to be communicated and agreed by all concerned and help promote effective teamwork.

Procedures covering the relationships and arrangements for record keeping, monitoring progress, time and cost control, risk management, project control and administration of the project should be developed, with the assistance of parties involved, for all stages of the project and cover time, costs, quality and reporting/decision-making arrangements.

The organisation structure should clearly identify the involvement and obligations of the client and his organisational backup.

Information and communication technology

Most construction projects employ electronic communications to a greater or lesser extent. Indeed, it would be almost unthinkable now not to use electronic devices to control, handle, manipulate and manage data, documents, information systems, and the supply chain without some form of electronic process.

The main purpose of information and communication technology (ICT) is to support efficient working practices and enable knowledge circulation. However, ICT systems have becoming increasingly more complicated in their operation and understanding and more demanding of high-specification hardware and specialist operator training with a need for (almost) continuous updating of all aspects (i.e. software, hardware and operator knowledge). The twin ideals of software that is both *intuitive and complex* is compromised when an apparently simple request which requires information to pass between computers, reveals that individual pieces of software will not 'talk' to each other. In a survey for the Institute of Directors and Dell Computers ('SMEs: Successful Growth through Innovation, IoD Business & Technology report 2007') it was found that data security and business continuity were cited as the most worrying issues for businesses in relation to ICT use. At the same time they also discovered that 95% of construction and manufacturing companies believed that chief executive officer leadership was essential for the initiation of major changes to ICT systems.

It is usual for extensive use to be made of computer applications as tools to assist most project management functions. It is essential for project managers to keep abreast of developments in this area in order to select and recommend appropriate packages and communication protocols for use on a project. It is particularly important to make sure that systems used by project team members are compatible to facilitate electronic exchange of data. Email, project specific websites, integrated project data applications, extranet applications and videoconferencing are examples of tools which may be required and the project manager will need to be able to define the ways in which these tools are used and how the transfer of information is managed and monitored. Examples and further guidance towards successfully utilising ICT in construction projects is available from IT Construction Best Practice (ITCBP) and IT Construction Forum publications. For some brief guidance on project management software see Appendix 12.

Building information modelling

First introduced in the 1970s this has become more than just the use of architecture, engineering and construction (AEC) design packages to design buildings and generate three-dimensional images and replicate visual attributes, e.g. colours and textures. Higher level integrated systems are now capable of designing in four dimensions (time dimension added) creating simulations running in real-time. Further work is being done to develop '*n*D' holistic modelling systems (i.e. *n* dimensions – multidimensional) having multiple data objects, e.g. drawings, schedules, graphical and non-graphical documents, and schedules. Such is the efficiency of these ICT systems that changes to any one aspect of any document can immediately be 'seen' by other users who in turn can act on them, thus resolving the perennial communication problems once slowing down projects.

The reserved and conservative nature of the construction industry causes it to monitor then adopt the good practices of other industries slowly. The use of integrated ICT systems such as those used by automobile and aerospace manufacturers is beginning to happen. The changes in operational procedures arising from the critical appraisal and adaptation of their practices for use by the construction industry are now emerging. Construction projects are often viewed as being 'unique' and this is often cited as a main factor when projects run over time and budget. However, it must be seen that critical examination of the construction manufacturing process and comparisons between the complex supply chains of the automobile and aerospace manufacturers are closer in alignment than has been acknowledged in the past. The adaptation and extension of these systems will need careful design if they are to become sustainable core elements of the management of projects. As before, the ability of such systems to intercommunicate is paramount if the outcome of construction projects is to become more predictable and client satisfying. Further introductory information regarding building information modelling is given in Appendix 29.

Project planning

The project master programme should be developed and agreed with the client and the consultants concerned, and detailed schedules for each stage of the project should be prepared as soon as the necessary parameters are established (see Appendix 3).

While preparing the master programme, necessary allowances should be incorporated to provide for potential delays (including the possible impact on initial revenue generation) in activities such as applying and obtaining statutory approvals, external consultations and enquiries, legal and funding negotiations and any other third-party agreements.

It is the project manager's responsibility to monitor the progress of the project against the master and stage programmes, identify risks to progress and to initiate necessary action to rectify potential or actual non-compliance.

For additional guidance on project scheduling see the CIOB publication on time management for complex projects.

Cost planning

A development budget study is undertaken to determine the total costs and returns expected from the project. A cost plan is prepared to include all construction costs and all other items of project cost including professional fees and contingency. All costs included in the cost plan will also be included in the development budget in addition to the developer's returns and other extraneous items such as project insurance, surveys and agent's fees or other specialist advisors.

The objective of the cost plan is to allocate the budget to the main elements of the project to provide a basis for cost control. The terms *budget* and *cost plan* are often regarded as synonymous. However, the difference is that the *budget* is the limit of expenditure defined for the project, whereas the *cost plan* is the definition of what the money will be spent on and when. The cost plan should, therefore, include the best possible estimate of the cash flow for the project and should also set targets for the future running costs of the facility. The cost plan should cover all stages of the project and will be the essential reference against which the project costs are managed.

The method used to determine the budget will vary at different stages of the project, although the degree of certainty should increase as more project elements become better defined. The budget should be based on the client's business case and should change only if the business case changes. The aim of cost control is to produce the best possible building within the budget.

The cost plan provides the basis for a cash flow plan, based on the master schedule, allocating expenditure and income to each period of the client's financial year. The expenditures should be given at a stated base-date level and at out-turn levels based on a stated forecast of inflation. A cash flow histogram and cumulative expenditure graph are shown in Figure 3.4.

Operational cost targets should be established for the various categories of running costs associated with the facility. This should accompany the capital cost plan and be included in the brief to consultants. The importance of revenue, grants and tax planning for capital allowances must also be taken into consideration.

When the cost plan is in place it serves as the reference point for the monitoring and control of costs throughout a project. The list which follows should be used as an aid in setting up detailed cost-control procedures for all stages of a project.

Cost control

The objective of cost control is to manage the delivery of the project within the approved budget. Regular cost reporting will facilitate, at all times, the best possible estimate of:

- established project cost to date
- anticipated final cost of the project
- future cash flow

In addition, cost reporting may include assessments of:

- ongoing risks to costs
- costs in the use of the completed facility
- potential savings.

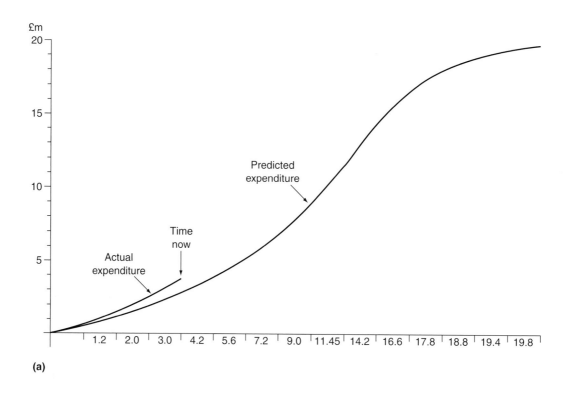

(a)

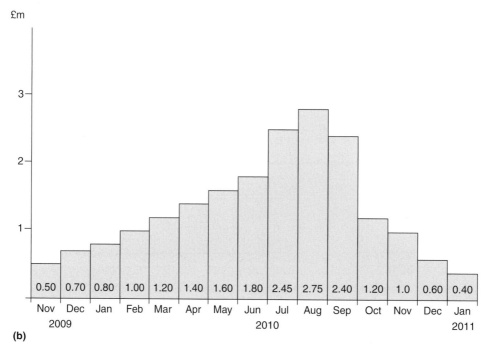

(b)

Figure 3.4 Examples of: (a) construction expenditure graph and (b) cash flow histogram.

Monitoring of expenditure to any particular date does not exert any control over future expenditure and, therefore, the final cost of the project. Effective cost control is effectively achieved when the whole of the project team has the correct attitude to cost control, i.e. one which will enable fulfilment of the client's objectives.

Effective cost control will require the following actions to be taken:

■ Establishing that all decisions taken during design and construction are based on a forecast of the cost implications of the alternatives being considered, and that no decisions are taken whose cost implications would cause the total budget to be exceeded.

■ Encouraging the project team to design within the cost plan, at all stages, and adopt the variation/change and design development control procedures for the project. It is generally acknowledged that 80% of cost is determined by design and 20% by construction. It is important that the project team is aware that no member of the team has the authority to increase costs on its section or element of the work. Savings on another must always balance increased costs on one item.

■ Regularly updating and reissuing the cost plan and variation orders causing any alterations to the brief.

■ Adjusting the cash flow plan resulting from alterations in the target cost, the master schedule or the forecast of inflation.

■ Developing the cost plan in liaison with the project team as design and construction progresses. At all times it should comprise the best possible estimate of the final cost of the project and of the future cash flow. Adherence to design freezes will aid cost control. (Development also means adding detail as more information about the work is assembled, replacing cost forecasts with more accurate ones or actual costs whenever better information can be obtained.)

■ As part of risk management, reviewing contingency and risk allowances at intervals and reporting the assessments is essential. Development of the cost plan should not involve increasing the total cost.

■ Checking that the agreed change management process is strictly followed at all stages of the project is very important (see Appendix 13). (It should be only during the construction phase of the project, when it can be demonstrated that significant delay, cost or danger would have been incurred by awaiting responses that the procedure should be carried out retrospectively.)

■ Arranging that the contractor is given the correct information at the correct time in order to minimise claims. Any anticipated or expected claims should be reported to the client and included in the regular cost reports.

■ Contingency money based on a thorough evaluation of the risks is available to pay for events which are unforeseen and unforeseeable. It should not be used to cover changes in the specification or in the client's requirements or for variations resulting from errors or omissions. Should the consultants consider that there is no alternative but to exceed the budget, a written request to the client must be submitted and correct authorisation received. This must include the following:

　○　details of variations leading to the request

　○　confirmation that the variations are essential

　○　confirmation that compensating savings are not possible without having an unacceptable effect on the quality or function of the completed project.

■ Submitting regular, up-to-date and accurate cost reports to keep the client well informed of the current budgetary and cost situation.

■ Establishing that all parties are clear about the meaning of each entry in the cost report. No data should be incorrectly entered into the budget report or any incorrect deductions made from them.

■ Ensuring that the project costs are always reported against the original approved budget. Any subsequent variations to the budget must be clearly indicated in the cost reports.

■ Plotting actual expenditure against predicted to give an indication of the project's progress (see Figure 3.4).

Procurement

In the context of this *Code of Practice*, procurement should be considered to be the process of identification, selection and commissioning of the contributions required for the construction phase of the project. The alternative methods of procurement referred to reflect the different organisational and contractual arrangements which can be made to ensure that the appropriate contributions are properly commissioned and that the interests of the client are safeguarded.

The various procurement options available reflect fundamental differences in the allocation of risk and responsibility to match the characteristics of different projects, therefore selection of the procurement option must be given strategic consideration. The project manager should advise on the relative benefits and disadvantages of each option, related to the particular circumstances of the project, for the benefit of the client.

The final choice of procurement method should be made on the basis of the characteristics of the project, the client and his requirements. The selection of method should be made when consideration is being given to the appointment of design and other specialist consultants because each option can have a different impact on the terms of appointment of the members of the project team.

The various procurement methods which may be pursued can be broadly classified under four headings:

■ traditional

■ design and build

■ management contracting

■ construction management.

Each method has its own variations. No method is best in all circumstances. They bring different degrees of certainty and risk towards the project construction and development.

Traditional

The contractor builds a client-designed scope of work within a time period for a lump sum. The client remains responsible for the design and the performance of his consultants under the building contract. The client appoints a design team, including a quantity surveyor responsible for financial and contractual advice. A building contractor is appointed, usually after a tender process, and often based on one of the standard forms of contract, to carry out the construction. The tender process can be based on complete design information or partial design information plus provisional guidance if an early construction start is required.

Some traditional contracts also allow for design responsibility for certain elements of the works to be passed to the main contractor, sometimes referred to as contractor's design portions (CDPs). This is particularly relevant where there are specialist elements of the works that need to be designed and installed by a specialist sub-contractor as part of the main contractors works.

Design and build

The client appoints a building contractor, usually on a standard form of contract to carry out at least some of the design and to complete construction within a time period for a lump sum. The contractor is responsible for the design elements and construction as defined in formal documentation known as the client's requirements, which usually impose the same design responsibility as would be imposed on a designing consultant, that of 'due skill and care'; it does not normally require the contractor to warrant 'fitness for purpose'. The appointment may be made after a tendering process incorporating variations on the method, or through negotiation. The client may appoint consultants to prepare the client's requirements, which may involve varying degrees of design and after-contract to oversee matters on his behalf. Difficulties can arise in distinguishing changes in the client's requirements (for which the client is liable) from design development (for which the contractor is liable).

If the design-and-build contractor is appointed after completion of a part of the design, the appointments of the design team may be formally passed on (contractually novated) to the design-and-build contractor. The purpose of the novation is to secure that what was the client's designer can maintain the required intent and quality throughout the technical design and production information stage when working under the direction of the contractor. However, in practice, research has shown the potential for conflicts and poor quality often remains.

Prime contracting is an extension of the design-and-build concept. The prime contractor will be expected to have a well-established relationship with a supply chain of reliable suppliers. The prime contractor co-ordinates and project manages throughout the design and construction period to provide a facility which is fit for the specified purpose, and meets its predicted through-life costs. The prime contractor is paid all actual costs plus profit incurred in respect of measured work and design fees; it is only at risk in respect of its staff and preliminaries.

Management contracting

The client appoints a design team with responsibilities, as in the traditional method, and augmented by a management contractor whose expertise and advice is available throughout the design development and procurement processes. Specialist works subcontractors, who are contracted to the management contractor on terms approved by the contract administrator who may be the architect, the quantity surveyor or the project manager, carry out the construction. The appointments of the management contractor and the trade subcontractors are usually made on standard contract forms. The management contractor is reimbursed all his costs and paid a percentage on project costs in the form of a guaranteed profit or fee.

Construction management

Construction management requires that the specialist works contractors are contracted to the client directly, involving the construction manager (which can and often is most successful as the leader of the design team) as a member of the project team acting as an agent and not a principal, to concentrate on the organisation and management of the construction operations. The project team, including the construction manager, is responsible for all financial administration associated with the works. The construction manager is paid an agreed fee to cover the costs of staff and overheads. This is generally considered to be the least adversarial form of contract and is often invoked when design needs to run in parallel with construction. This method is at its best with a hands-on responsive client who can make decisions quickly. On the other hand, it has been found to relatively unsuccessful when the client needs to refer issues to one or more committees before decision.

Relevant issues

Variations from the formats described above can be a potential source of confusion and compromise the intended philosophies. Before a contractual or organisational variation is introduced the choice of procurement option should be made against the most important criteria. Only then should essential variations be introduced and these must be dealt with by specific contractual arrangements and documentation within the framework of the overall procurement method adopted. Further guidance on alternative forms of procurement has been included in Appendix 28.

It is important to recognise that the actual process by which construction projects are implemented is project specific, and given a particular project, will remain the same irrespective of the procurement route is followed. This process involves four stages:

1. Development of a detailed definition of the requirements for the producer (the project brief).

2. Preparation of designs, working drawings and specifications identifying every component and detailing the construction method.

3. Procurement of every component required for the product, and the specialised skills necessary for its construction.

4. Construction and commissioning of the product.

Characteristics of alternative procurement options

The characteristics of four basic forms of procurement are explained in detail in Appendix 15. In addition further guidance relating to procurement of project team partners through framework contracts have been provided in Appendix 22.

Appointment of project team

The project manager in consultation with the client will decide on and implement a selection procedure for members of the protect team, together with contractors and other consultants. These may then be appointed on behalf of the client. The extent of contractors and consultants will be determined, to some extent, by the procurement route selected. In connection with the project team, two common arrangements for appointment are:

■   separate appointment of independent service providers

■   single appointment of a team of service providers or a lead organisation for the provision of all services.

It is important that the members of the project team should be as compatible as possible, both in temperament and in working methods if the project is to have the greatest likelihood of success. Therefore, selection should be based on balancing quality, compatibility, schedule and price (see *Selecting Consultants for the Team: Balancing Quality and Price*, CIB/Thomas Telford, 2000).

The project team, consultants and contractors (who may or may not also be suppliers within the supply chain), may be appointed through a process of short-listing and structured interviews or through a competitive tendering procedure. This may be through European Union (EU) procurement procedures, which may be mandatory for publicly funded projects depending on the size of the project (see Appendix 5 for a brief guidance to EU procurement rules). The project manager needs to be fully informed on all issues related to the procurement process and advise the client accordingly.

The client should be consulted on the formulation of shortlists and should be invited to attend any interviews. The process is set out in Table 3.1.

Table 3.1 Appointment of the project team

Activity	Considerations
Selection and appointment of project manager	May be appointed at inception/feasibility stage
Agree criteria for team selection	Type of expertise and scope
	Budget fee
	Contractual procurement strategy
Define in detail each assignment	Extent of services required
	Co-ordinate with other professional agreements
Define roles and duties	Scope of work
	Roles and duties
Agree terms and conditions of engagement	Client's of standard conditions of engagement
	Programme, professional indemnity (PI) insurance, warranties
Selection of those invited to bid	Utilise relevant databases
	Agree list of consultants
	Decide selection criteria
	Format proposal required
	Agree content and fee
Carry out the selection process	Agree interview team
	Agree information to be used
	Arrange interviews
	Utilise scoring system
Negotiate conditions of appointment	Provide client with analysis and selection recommendations
	Negotiate final conditions
Advise on final appointment	Issue letter of appointment
	Issue letters of rejection
Oversee formalities relating to PI insurance, warranties and building defects insurance	Legal department to formalise
	Finance department for fees
	Legal documents issued

For the 'short-list' method, the project manager should formulate the short lists, convene and chair the interviews, record and assess the results and present a report and recommendation to the client for final decision (see Appendix 14 of Part 1 and Appendix B of Part 2 for further guidance on the selection process).

Most professional firms are members of organisations that publish standard terms of appointment and codes of conduct. It is usual to appoint the project team members on the standard terms which are designed to provide a proper balance of risk and responsibility between the parties. Standard terms are capable of amendment by agreement (but any amendment should be considered very carefully before changes are made, a common practice is for solicitors to amend clauses inconsistently), but the project manager should advise against terms which impose uninsurable risks or unquantifiable costs on the consultants or are in conflict with their professional responsibilities or codes of conduct.

The project manager will issue to the appointed project team with the project handbook, the outline project brief and the master schedule together with the budget or cost plan. It is advisable that these elements are referred to in as much detail as is available at the time of the project team's appointment.

Procuring the supply chain

When selecting the key members of the supply chain, not only should the key contractors be appointed as soon as is feasible, depending on the procurement route selected, but key suppliers, especially those who may have a design element within their requirement, should likewise be appointed at an early stage of the project. This should enable the following to be achieved:

■ resolution of the buildability issues at design stage

■ choice of the most efficient materials to be used

■ advice to the client as to costing (practical, rather than rates)

■ co-ordination of the 'designer suppliers' with main design team

■ understanding of the client's needs in all areas, which will assist in the construction process leading to the quality and required product for the client

■ highlighting and incorporation of relevant health and safety issues which must be taken into account on the design.

Tender procedure

The outline development schedule will indicate the time allowed for procurement and design, will show activities such as tender interviews, tendering and selection; scope, release dates, approval periods, cost checking and appropriate documentation, etc. These activities may include the following (see also Figure 3.5):

■ Checking that the various documents are produced at the appropriate times, including those for pre-main construction works (e.g. demolition, site clearance, access and hoarding) and ensuring that they contain any special terms required by the client and local statutory authorities for activities such as archaeology and environmental investigations. In conjunction with the relevant consultants, preparing lists of firms to be invited to tender for the elements of the work (pre-qualifying process). Obtaining confirmation that the listed firms will be prepared to submit their offers at specified dates, taking up references and/or interviewing prospective supply chain companies, including relevant consultants.

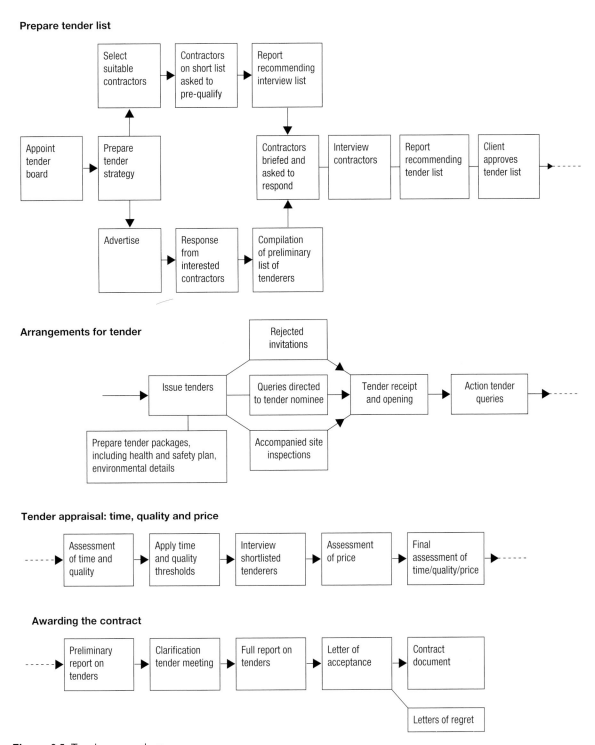

Figure 3.5 Tender procedure.

- ■ Ensuring that all appropriate health and safety/CDM references are made in the documentation together with any relevant statutory authority references.

- ■ Checking, in liaison with other project team members, that all contract documentation is co-ordinated, especially where aspects of design are involved and confirming that, where appropriate, warranties are secured.

- ■ Receiving reports on supply chain procurement, together with method statements.

- Interviewing successful supply chain applicants, if necessary, to clarify any special conditions and to meet significant leading personnel.

- Arranging for formal acceptance of successful supply chain applicants as appropriate and issuing relevant letters of intent.

- Selection should be based on balancing quality, schedule and price.

- Initiating action if price submissions are outside of the budget.

- Ensuring that the client understands the nature and terms of the construction, particularly those in relation to possession and payment terms, and that possession of the site can be given to the contractor on the date set out in the tender.

- Arranging for formal signing and exchange of contracts.

In view of the EU directives on procurement, negotiated tendering is also being undertaken as an option to secure best value for money. See Appendix 5 for further information about the guidance on EU procurement rules.

Partnering

Partnering is a set of actions by project team by which risks can be distributed and conflict can be minimised. The intention is to provide 'win-win' outcomes for everyone involved. It is put into practice by having regular partnering workshops of all the key members of the project team to establish and foster co-operative ways of working aimed at improving performance. In broad terms partnering teams agree mutual objectives that take account of the interests of all the parties; establish co-operative methods of decision-making including procedures for resolving problems quickly; and identify actions to achieve specific improvements to normal performance. The workshops take place throughout the project usually under the guidance of an independent partnering facilitator.

There is considerable evidence to show that partnering can bring significant net benefits in the form of reduced costs, improved quality and shortened timetables. However, these benefits involve additional costs in ensuring the selected project team members are willing to work co-operatively and the expense of running partnering workshops. The additional costs arise during the early stages of projects while the benefits, which depend on partnering being successful, accrue during the later stages. The likely costs and benefits of partnering should be considered by all clients. It is very likely to provide benefits for clients with phased projects or who undertake schedules of similar projects, but it has been shown to provide benefits on a one-off project. Partnering of the supply chains that produce key elements should also be considered on all except the smallest and simplest projects (see Appendix 8). CIOB's *Partnering in the Construction Industry: Code of Practice for Strategic Collaborative Working* provides detailed guidance and advice.

Procurement under EU Directives

EU Directives (Directive 2004/17/EC and 2004/18/EC dated 31 March 2004) enacted by UK (The Utilities Contracts Regulations 2006 and The Public Contracts Regulations 2006 enforced 31 January 2006) require that for public procurement above a certain monetary threshold, contracts must be advertised in the *Official Journal of the European Union* (OJEU) and there are other detailed rules that must be followed. This is irrespective of the procurement route chosen. Further advice and guidance on this has been provided in Appendix 5.

e-Procurement

Electronic procurement (e-procurement) is described to be the use of electronic tools and systems to increase efficiency and reduce costs during procurement. The EU Consolidated Directives and EU Invoicing Directives provide clear directions for enabling e-procurement, including encouraging online-only processes whereby tenderers can compete for contracts. In addition to key public sector clients such as the Ministry of Defence, the National Health Service and local authorities, private sector clients are also increasingly utilising this process of procurement. Further guidance on e-procurement has been provided in Appendix 23.

Framework agreements

Rather than uniquely tender goods or services on individual projects, some large clients prefer to establish framework agreements with preferred suppliers. Frameworks can be described as agreements to provide both goods and services on predefined and specified terms and conditions with a selected number of suppliers (e.g. consultants, designers and contractors). Frameworks range from a mechanism to assemble a limited number of prequalified suppliers who will then be asked to bid for projects to arrangements where specific teams of suppliers are guaranteed a regular workload in return for continuous improvement in delivery of their product. There are a wide variety of framework procurement contracts available for use. Brief guidance on procurement and management of framework contracts has been provided in Appendix 22.

4 Pre-construction stage

Client's objectives

At this stage the client expects to finalise the project brief for the project team, identify and agree the solution that gives optimum value, and to ensure a technical design which can be efficiently delivered with predictability of cost, time and quality.

Interlinking with previous stages

After the client has made a commitment to the project, accepted the feasibility report and approved the concept design, the process will then move into the next phase or stage which we call pre-construction. However, it should be appreciated that on a number of projects, many of the stages overlap and it is only in order to identify the full scope of activities involved in the development process and to enable some sort of chronology to be established, that we have sequenced the activities into stages.

'Pre-construction' involves establishing the technical design, the preparation of tender documents, the tendering process (including negotiated tendering). However, the precise sequence of activities will depend very much on the choice of procurement system, and the type and form of contract selected.

It is worth noting at this stage, that we are moving into an ever-increasing legislative environment, with greater controls in the form of statutory requirements, national and European legislation and guidelines, minority stakeholder pressures, demand for greater sustainability and growing restrictions on disposal of unusable material, to name but a few. Therefore, by the start of this pre-construction stage, a significant number of key activities will have been addressed and actioned. These include the following:

■ The client's project brief detailing the anticipated objectives for the project will have been established and the associated concept design fundamentally completed. However, while it would be hoped that the design brief would remain substantially unchanged for the reminder of the project, it is possible that unforeseen factors could have some effect on the brief during the project period, although hopefully these will be minimal.

■ A suitable site and the scope of any treatments required, will have been identified and made available.

■ Environmental and energy audits will have been undertaken.

■ A risk register will have been prepared incorporating data from risk analysis.

■ Surveys to cover geology, topography, hazardous materials (COMAH), landfill and recycling will have been carried out.

- Obligatory reports concerning sustainability, disability discrimination, etc., will have been prepared and approved by the appropriate authorities.

- Statutory requirements concerning the Construction Contracts Act (see Appendix 7) and CDM regulations will have been accommodated.

- Statutory authorities, public bodies and utilities will have been approached for information regarding all mains services, highways and related infrastructure items, which are likely to influence site development.

- A development schedule will have been prepared.

- A cost plan will have been prepared.

- A cost allowance will have been allocated to cover on-site development including pre-main construction works, infrastructure, buildings, fitting-out and equipment.

- The planning authorities will have been contacted regarding the planning status of the site, which has been deemed acceptable for the intended purpose. Outline planning consent will have been obtained.

- The project team will have been appointed together with their associated consultants. This team will include the client, the project manager and, as soon as possible, representatives from the main contractor and associated key subcontractors/work packages. These will all contribute to the strategic decision-making process.

- The project execution plan (PEP), drafted during the feasibility stage, may be enhanced during this stage. It is a live document which governs the strategy, organisation, control procedures, respective responsibilities for the project and much more:

  ○ client brief: functional and aesthetic; business plan

  ○ constraints and risk assessment; revenue assumption/criteria

  ○ funding cost controls: budget; drawdown procedures; time and cost contingencies

  ○ schedule: deadlines, milestones

  ○ organisation and resources: responsibilities, delegated authority

  ○ project strategy and procurement details

  ○ roles, responsibilities and outputs of project team members (*modus operandum*)

  ○ occupation plan: commissioning; facilities management/maintenance strategy

The project handbook would have been prepared under the guidance of the project manager and submitted to the client and any other interested party for comment, discussion and agreement. Its review and update will be the responsibility of the project manager, unlike the health and safety file which is the responsibility of the CDM co-ordinator. This project handbook differs from the PEP, in that the handbook sets out the process and procedures for administration purposes, whereas the PEP covers detail as shown in Chapter 2 and in the preceding bullet points.

The client will have authorised the project to proceed and should be aware that considerable costs will now begin to be incurred. Adequate cash flow provision must be provided for expenditure principally at regular monthly intervals. These will include professional services fees, e.g. for the project manager, architect, quantity surveyor, structural and M&E engineers, project planner/scheduler, together with planning fees and on-site investigations, demolition, site clearance and disposal, etc.

The pre-construction stage is about final preparation for the construction stage. The success of which will depend, to a great extent, on the amount of pre-planning and preparation that has taken place during this and earlier stages.

Managing the design delivery

The project manager will need to convene a meeting of the design team and any other consultants/advisors to review all aspects of the project to date. A dossier of relevant information should be circulated in advance. The object of the meeting will be to formulate a design management plan. The plan should at least cover:

■ who does what by when

■ the size and format of drawing types

■ schedules of drawings to be produced by each discipline/specialist

■ relationships of interdependent CAD (computer-aided design) systems

■ transfer of data by information technology

■ estimates of staff hours to be spent by designers on each element or drawing

■ monitoring of progress and the effect of design resources expended compared to productivity achieved

■ schedules of information required/release dates

■ initiating procedures for design changes to be made and their effects predicted

■ incorporation within the design schedule of key dates for review of design performance to check:

    ○ strategy and technical options to meet the sustainability brief

    ○ compliance with brief

    ○ cost acceptance

    ○ value engineering analysis

    ○ health and safety issues

    ○ completeness for tender.

The project manager is ultimately responsible for ensuring that there is a system in place for monitoring and controlling the production of design information in line with the agreed schedule. The project manager should convene and attend regular design team meetings to review progress and ensure that the design team are performing in accordance with their duties. Responsibility for the co-ordination and integration of the works, which involves input from other designers, consultants, service providers, statutory authorities and utilities, etc., normally would lie with the design team leader, which in most cases will be the architect.

Suggested task list for the design team leader

- Establish the overall design style, quality, etc.

- Establish a grid/reference system for the base scheme.

- Review the design schedule.

- Direct the design process.

- Liaise with the client about significant design issues.

- Prepare sufficient production information for consultants and specialists to develop their proposals, co-ordinating these and integrating them into the overall scheme.

- Advise on the need for and appointment of other consultant and specialists.

- Establish a system for information transfer; check compatibility of system and software.

- Co-ordinate the briefing document.

- Establish a system of design reviews and validation.

- Agree a basis for the cost plan to be developed and its subsequent monitoring.

- Undertake his role and duties under the CDM regulations: this would include ensuring that only 'initial design' is carried out until a CDM co-ordinator has been appointed.

Duties of the project manager at this stage

- Organise within the client organisation appropriate groups of people, who will contribute to the detail of the brief and champion relevant aspects of the design prepared by the design team for signing off.

- Assist in the preparation or finalisation of the project brief.

- Prepare the design management plan and liaison with the CDM co-ordinator to ensure compliance with the CDM regulations at this stage.

- Arrange the appointment of other consultants and specialists.

- Organise the communication and information systems.

- Produce and co-ordinate the design schedule and monitor progress and productivity.

- Ensure that various technical specialists appointed by the client such as IT, acoustics, catering, landscaping and artists are brought into the design process at the appropriate times.

Project co-ordination and progress meetings

To aid control of the design process, the project manager will arrange and convene project progress meetings at relevant intervals to review progress, resources and productivity on all aspects of the project and initiate action by appropriate parties to ensure that the design management plan is adhered to or establish reasons for departure and the effects of departure. In the event of a departure, the project manager should address contingency planning and mitigatory/recovery strategy and initiate recovery. Distributing minutes of meetings to all concerned is an essential part of the follow-up action.

Design team meetings

Design team meetings are convened, chaired and minuted by the design team leader. It is not essential for the project manager to attend all team meetings as a matter of course, although he normally has the right to do so. The project manager will receive minutes of all meetings and will report to the client accordingly.

Managing design team activities

Key specialist contractors may need to be involved at an early stage and managed equally with the design team (see Figure 4.1).

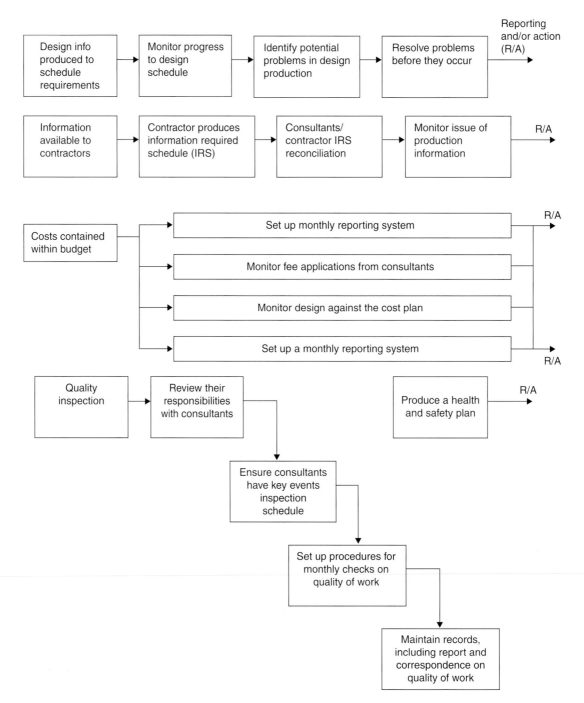

Figure 4.1 Design team activities.

The project manager has several responsibilities:

■ Monitoring progress, resources and productivity against the design management plan in association with the team. This is essential in view of their interrelationship. However, effective interrelationship cannot be finalised until the full team has been appointed and had time to get to grips with the project and its complexities.

■ Advising the design team leader of the requirement to agree the detail and integration of the design team activities and to submit an integrated design production schedule for co-ordination by the project manager.

■ Incorporating into the project schedule the dates for the submission of design reports and periods for their consideration and approval.

■ Commissioning, as necessary, or arranging for the team to commission, specialist reports, e.g. relating to the site, legal opinions on easements and restrictions and similar matters.

■ Ensure a competent consultant is appointed as CDM co-ordinator as required by CDM regulations.

■ Drawing to the attention of the client and the designers their respective duties under the CDM Regulations and monitoring compliance.

■ Arranging for the team to be provided with all the information it requires from the client in order to execute its duties. It is an important function of the project manager to co-ordinate the activities of the various (and sometimes numerous) participants in the total process. CDC co-ordinator, solicitors, accountants, tax advisors, development advisors, insurance brokers and others may all be involved in the pre-construction stage.

■ Submitting, in conjunction with the design team leader, design proposals, reports and design development (formerly scheme design) drawings to the client for approval (see Figure 4.2).

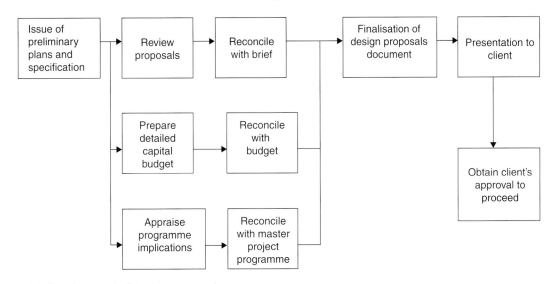

Figure 4.2 Development of design proposals.

■ Conveying approvals to the team to proceed to subsequent stages of the project.

■ Obtaining regular financial/cost reports and monitoring against budget/cost plans. Initiating remedial action within the agreed brief if the cost reports show that the budget is likely to be exceeded. Solutions to problems that cannot be resolved within the agreed brief, or likely substantial budget underspend, should be submitted to the client with recommendations. The necessity to agree firm budgets at an early stage is most essential. It could, in certain cases, lead to the client modifying the project brief.

■ Preparing 'schedule of consents' with action dates submission documents, status, etc., and monitor progress.

■ Checking that professional indemnity insurance policies are in place and remain renewed on terms that accord with conditions of engagement.

Statutory consents

Although a great deal of the detailed work involved in obtaining statutory consents, such as planning permission and Building Regulations' approval, is carried out by the design team and other consultants, the project manager has a vital facilitating role to play in what can be critical project activities.

Planning approval

Planning consultants
On many schemes today the complexity of the planning process is such that the project manager may appoint a planning consultant to advise on the approach most likely to secure planning consent for a scheme together with the information that will be required to support the application, including an environmental impact assessment incorporating sustainability appraisal, traffic assessment, green travel plan and ecology report. In conjunction with the other members of the design team, the planning consultant will organise and participate in key meetings with the planning officer and other departments, such as highways.

Legislation
The primary legislation governing the planning process is contained in several Acts of Parliament. The grant of planning does not remove the need to obtain any other consents that may be necessary, nor does it imply that such consents will necessarily be forthcoming.

Planning permission
On many schemes today the complexity of the planning process is such that the project manager may appoint a planning consultant to advise on the approach most likely to secure planning consent for a scheme together with the information that will be required to support the application, including a sustainability impact assessment, traffic assessment, green travel plan and an ecology report.

In conjunction with other members of the design team, the planning consultant will organise and participate in key meetings with the planning officer and other departments, such as highways.

Timing
Planning permission cannot be guaranteed or assured in advance of the local planning authority (LPA) decision within the statutory period and the project manager must recognise this by allowing a contingency in the outline development schedule.

Negotiations
The project manager will normally assist the design team leader in negotiations with officers of the local authority and report to the client on the implications of any special conditions, or on the need to provide *planning gain* through the appropriate statutory agreements. The client's legal advisors are briefed to act for the client accordingly.

Presentations   The project manager will arrange, should it be necessary, any presentations to be made to LPAs and local community groups. He will also organise meetings, including agreeing publicity and press releases with the client.

Refusal   Should planning permission be refused the advice of the relevant consultants should be obtained and action initiated, either to submit amended proposals or to appeal the decision.

Appeal   In the event of an appeal, arrangements are made for the appointment and briefing of specialists and lawyers, including managing the progress of the appeal. Applicants who are refused planning permission by an LPA, or who are granted permission subject to conditions which they find unacceptable, or who do not have their applications determined within the appropriate period, may appeal to the Secretary of State. Appeals are sent to the Planning Inspectorate.

Enforcement powers   In the event of a breach of the planning legislation, the local authority's main enforcement powers are:

- to issue an enforcement notice, stating the required steps to remedy an alleged breach within a time limit (there is a right of appeal to the Secretary of State against a notice)

- to serve a stop notice which can prohibit, almost immediately any activity to which the accompanying enforcement notice relates (there is no right of appeal to the Secretary of State)

- to serve a breach condition notice if there is a failure to comply with a condition imposed on a grant of planning permissions

- to apply to the High Court or County Court for an injunction to restrain an actual or apprehended breach of planning control

- to enter privately owned land for enforcement purposes; and

- following the landowner's default, to enter land and carry out the remedial work required by an enforcement notice, and to charge the owner for the costs incurred in doing so.

It is a criminal offence not to comply with an enforcement notice's requirements or to contravene the prohibition in a stop notice.

Other statutory consents

It is the duty of the design team to facilitate that the design complies with all other statutory controls, e.g. consents for Building Regulations, means of escape, the storage of hazardous materials, fumes and emissions, and pollutants. Generally, statutory controls make the owner or occupier responsible for the aspect of continuing duties in relation to the statute. The project manager obtains from the design team and/or other relevant sources, all consents and arranges for the client to be advised of these continuing duties. Others, such as specialist subcontractors, submit and obtain Building Regulations' approval for their product/system.

Impact of utilities on project planning/scheduling

Due to the long lead periods required by utilities providers, i.e. water, gas, electricity, the project manager should ensure that the requirements in respect of diverting or increasing existing supplies or installing new supplies are identified at an early stage in the development process. The procurement of these utilities should be monitored closely to ensure that they do not affect the completion of the project.

The project manager should also be aware that supplies are often required prior to construction completion to enable commissioning of the building services installations.

Technical design and production information

The project manager's monitoring and co-ordinating role will entail extensive liaison with members of the project team and will include the tasks shown in Figure 4.3, which are set out in more detail below.

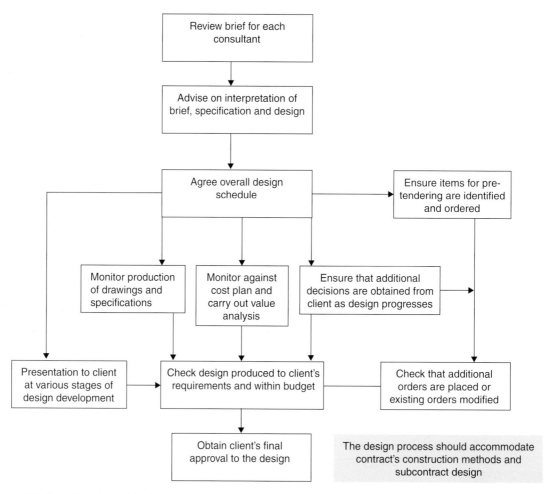

Figure 4.3 Co-ordination of design work up to design freeze.

- ■ Controlling the extent to which the design will be produced by specialist contractors and/or component manufacturers, and establishing the division of responsibilities between them and the design team.

- ■ Reviewing the project strategy, control systems, procedures and amending the project handbook, as required.

- ■ Amplifying the design brief as necessary during design development.

- ■ In conjunction with the project team, prepare updates to the outline development schedule for the detailed design and production information stage, defining tasks and allocating responsibilities.

- Updating the schedule to establish timely flow of information from the design team for:
  - cost checking
  - client's approval
  - tender preparations
  - construction processes.

- Co-ordinating the activities of the client and the project team in the management of the production of the design information.

- Formulating, in collaboration with the consultants, recommendations to the client/owner in respect of the quality control system, including:
  - on-site and off-site inspection of work for compliance with specifications, and testing of materials and workmanship
  - performance testing and the criteria to be used
  - updating the schedule to incorporate requirements for samples and mock-ups, their updating and monitoring progress of approvals; copy of the schedule is to be included in the relevant monthly reports.

- Listing the key criteria in terms of performance benchmarking that in all areas of design make clear how the design will be judged, i.e. air changes or faults with current facilities.

- Monitoring the emerging technical design against the risk register, cost plan and the development schedule.

- Liaising with the client and project team and the local authority and utility companies and other statutory bodies to obtain permissions and approvals.

- Evaluating changes in the client's requirements for cost and time implications and incorporating approved items into the design process.

- Reviewing progress, updating the development schedule and providing regular reports incorporating information relating to:
  - project status
  - intervening events and their effects
  - cost against budget/cost plan, together with reconciliation statement
  - forecast of total cost and date of completion
  - risk register and contingency planning
  - mitigation and recovery plan.

- Obtaining the client's approval to the detailed design and production information phase.

- Co-ordinate and liaise with the CDM co-ordinator to ensure arrangements for health and safety of planning and design work to enable satisfaction of the risk management hierarchy (Eliminate – Reduce – Inform – Communicate).

- Initiating arrangements for implementation of approved design and production information, to ensure that contractors' reasonable information requirements are fulfilled.

Pre-start meeting

The pre-start meeting with contractors and consultants (project team) is held to establish proper working arrangements, roles and responsibilities, lines of communication and agree procedures to be followed throughout the contract (project on-site). If bonds are required they must be provided before possession of site is granted. The 'principal contractor's' construction phase health and safety plan must be in place before work starts on the site. See Table 4.1 for a specimen agenda for a pre-start meeting.

Agenda items at pre-start meeting

Introduction

■ Introduce the representatives who will regularly attend progress meetings and clarify their roles and responsibilities. The client, contractor and consultants may wish to introduce themselves.

■ Briefly describe the project and its priorities and objectives, and any separate contract that may be relevant (preliminary, client's own contractors, etc.).

■ Indicate any specialists appointed by the client, e.g. for quality control, commissioning, for this contract.

Contract

■ Describe the position with regard to preparation and signature of documents.

■ Hand over any outstanding production information, including nomination instructions, variation instructions. Review situation for issuing other important information.

■ Request that insurance documents be available for inspection immediately, remind the contractor to check specialist subcontractors' indemnities. Check if further instructions are needed for special cover.

■ Confirm the existence, status and use of the information release schedule, if used. Establish a procedure for agreeing adjustments to the schedule should they be necessary.

Contractors' matters

■ Check that the contractors' working schedule is in the form required and that it satisfactorily accommodates the specialist subcontractors. It must:

○ contain adequate separate work elements to measure their progress and integration with services installations

○ allocate specific dates for specialist subcontract works, including supply of information, site operations, testing and commissioning

○ accommodate public utilities, etc.

■ Agree a procedure for the contractor to inform the architect of information required in addition to any shown on the information release schedule. This is likely to involve the contractor's schedule of information required, which must relate to his works' schedule and must be kept up to date and regularly reviewed. It should include information, data, drawings, etc. to be supplied by the contractor/specialist subcontractors to the design team.

■ Review in detail the particular provisions in the contract concerning site access, organisation, facilities, restrictions, services, etc. to ensure that no queries remain outstanding. Ensure that the contractor has a copy of any conditions placed on the client in respect of the planning consent. Also provide the contractor with legal drawings showing the curtilage of the site ownership.

Table 4.1 Specimen agenda for pre-start meeting

1. **Introductions**
   Appointments, personal
   Roles and responsibilities
   Project description

2. **Contract**
   Priorities
   Handover of production information
   Commencement and completion dates
   Insurances
   Bonds (if applicable)
   Standards and quality

3. **Contractors' matters**
   Possession
   Schedule
   Health and safety files and plan
   Site organisations, facilities and planning
   Security and protection
   Site restrictions
   Contractor's quality control policy and procedures
   Subcontractors and suppliers
   Statutory undertakers
   Overhead and underground services
   Temporary services
   Signboards

4. **Resident engineer/architect/clerk of works' matters**
   Roles and duties
   Facilities
   Liaison
   Instructions

5. **Consultants' matters**
   Structural
   Mechanical
   Electrical
   Others

6. **Quantity surveyor's matters**
   Adjustments to tender figures
   Valuation procedures
   Remeasurement
   VAT

7. **Communications and procedures**
   Information requirements
   Distribution of information
   Valid instructions
   Lines of communication
   Dealing with queries
   Building control notices
   Notices to adjoining owners and occupiers

8. **Meetings**
   Pattern and proceedings
   Status of minutes
   Distribution of minutes

- Quality control is the contractor's responsibility. Remind the contractor of the contractual duty to supervise standards and quality of work during the execution of the works.

- Determine the information that will require to be provided by the contractor to meet the sustainability strategy such as timber certificates and reports on waste management.

- Numerous other matters may need special coverage, e.g.:

    ○ check whether immediate action may be needed by the contractor over specialist subcontractors and suppliers

    ○ emphasise that drawings, data, etc. received from contractor or specialist subcontractors which are yet to be approved will remain the responsibility of the originator until approval

    ○ review outstanding requirements for information to or from the contractor in connection with specialist works

    ○ clarify that the contractor is responsible for co-ordinating performance of specialist works and for their workmanship and materials, and for co-ordinating site dimensions and tolerances.

- The contractor must also provide competent testing and commissioning of services as set out in the contract documents, and should be reminded that the time allocated for commissioning is not a contingency period for the main contract works.

- The contractor must obtain written consent before subletting any work.

Resident engineer/clerk of works' matters

- Clarify that inspections are periodic visits to meet the contractor's supervisory staff, plus spot visits.

- Explain the supportive nature of the various roles and the need for co-operation to enable the clerk of works and resident engineer to carry out their duties.

- Remind the contractor that the resident staff must be provided with adequate facilities and access, together with information about site staff, equipment and operations.

- Confirm procedures for checking quality control, e.g. through:

    ○ certificates, vouchers, etc. as required

    ○ sample material to be submitted

    ○ samples of workmanship to be submitted prior to work commencing

    ○ test procedures set out in the bills of quantities

    ○ adequate protection and storage

    ○ visits to suppliers' and manufacturers' works.

Consultants' matters

- Emphasis that consultants will liaise with specialist subcontractors only through the contractor. Instructions are to be issued only by the architect/contract administrator. The contractor is responsible for managing and co-ordinating specialist subcontractors.

- Establish working arrangements for specialists' drawings and data for evaluation (especially services) to suitable timetables. Aim to agree procedures which will speed up the process; this sector of work frequently causes serious delay or disruption.

Quantity surveyor's matters

■ Agree procedures for valuations; these may have to meet particular dates set by the client to ensure that certificates can be honoured.

Clarify:

■ the process and procedures to deal with foreseeable and unforeseeable changes

■ tax procedure concerning VAT and 'contractor' status

■ status of precedence among the construction issue specifications, drawings and bills of quantities if appropriate.

Communications and procedures

■ The supply and flow of information will depend on the updated working schedule and will proceed smoothly if:

○ there is regular monitoring of the working schedule

○ requests for further information are made specifically in writing not by telephone

○ the design team responds quickly to queries

○ technical queries are raised with the clerk of works (if appointed) in the first instance

○ policy queries are directed to the architect/contract administrator

○ discrepancies are referred to the architect/contract administrator for resolution.

■ On receiving instructions, the contractor should check for discrepancies with existing documents; check that documents being used are current.

■ Information to or from specialist subcontractors or suppliers must be via the contractor.

■ All information issued by the design team should be via the appropriate forms, certificates, notifications, etc. The contractor should be encouraged to use standard formats and classifications.

■ All forms must show the distribution intended; agree numbers of copies of drawings and instructions required by all recipients.

■ Clarify that no instructions from the client or consultants have any contractual significance and should not be acted on by the contractor or any subcontractor but should immediately be referred to the contract administrator for decision; only written instructions from the contract administrator are to be actioned under the contract and all oral instructions must be confirmed in writing. Explain the relevant procedure under the contract. The contractor should promptly notify the contract administrator of any written confirmation outstanding.

■ Procedures for notices, application or claims of any kind are to be strictly in accordance with the terms of the contract; all such events should be raised immediately the relevant conditions occur or become evident.

■ It is advisable that a communication plan is agreed in advance so as to provide a clear direction to all parties involved, particularly in complex projects with multiple stakeholders. Further guidance on formulating a communication plan is provided in Appendix 25.

Meetings

Review format, procedures, timing, participants and objectives of the next stage:

- meetings, i.e. site (progress) meetings, policy/principal's meetings and contractor's production information meetings, and

- site inspections.

Fee payments

The project manager is responsible, as required in the conditions of engagement, for receiving fee accounts and invoices from consultants and others concerned with the project, checking for correctness and arranging for payment within the terms of the various appointments or contracts.

Quality management

It is the project manager's role to set up and implement an appropriate process to manage project quality. From the quality policy defined in the project brief, the development of a quality strategy should lead to a quality plan setting out the parameters for the designers and for the appointment of contractors. Quality control then becomes the responsibility of the contractor, subcontractors and suppliers operating within the agreed quality plan. The plan itself should establish the type and extent of independent quality auditing (particularly for off-site production of components) and the timing of inspections and procedures for 'signing off' completed work.

It is the responsibility of the design team and other relevant consultants to specify the goods, materials and services to be incorporated in the project, using the relevant British Standards, codes of practice and **Board of Agrément criteria** or other appropriate standards.

The achievement of these standards rests with the appointed main contractor. When interviewing contractors at the pre-tender stage, the project manager will seek confirmation that each company has a positive and proactive policy towards the control of quality, a policy which will be reflected in all of its operations off or on the site.

Dispute resolution

The Woolf reforms recommended that litigation should be seen as a last resort and alternative dispute resolution such as mediation and other methods of resolving disputes are actively encouraged; parties who do not participate in mediation must justify their position to a judge.

Adjudication may be used to resolve disputes using the provisions of the Housing Grants, Construction and Regeneration Act, 1996 as amended 2009 (see Appendix 7).

Although it is hoped that the non-adversarial approach and the increasing choice of alternative procurement options and partnering will lead to a reduction in disputes, nevertheless, the project manager should make every effort to pre-empt any dispute that may arise and endeavour to mitigate and resolve the problem.

Arbitration is a procedure that may be used following adjudication and before initiating litigation. An outline of procedures to be applied in resolving contractual disputes is given in Appendix 16.

Avoiding common project management pitfalls

Particular attention and consideration is necessary to avoid the following:

- lack of clear links between project and strategic priorities including agreed measures of success

- lack of support, ownership and leadership

- lack of effective stakeholder engagement

- lack of effective project management and risk-management skills

- disproportionate sequencing and scheduling of activities

- too much emphasis on initial price rather than long-term value for money

- lack of understanding of (and communication with) the supply chain

- lack of integration between clients, the project team and the supply chain.

Good practice in project management is considered further in Appendix 26.

Contractual arrangements

The project manager has to ensure that all statutory and contractual formalities are in place prior to allowing work to start on site. It may mean that the project manager has to ensure that others have given the relevant notice and, if appropriate, received the relevant approval. A log may help to keep track of notices and approvals together with the owner of the task.

These are likely to include:

- planning consent

- third-party agreements (such as landlord's approval, party wall and rights of light)

- CDM notification

- insurances (such as professional indemnity, employer's liability, project insurance and third party)

- notice to start work under the Building Regulations

- Fire Regulation compliance

- performance bonds.

On completion various completion certificates are required, these should be specified in the particular specification, and would include:

- Fire Regulation compliance

- electrical completion certificate

- test certificates both manufacturing and installation

- lifting beams tests and marking

- Building Regulation compliance

■ pressure vessel and boiler certificates.

For special buildings or processes, e.g. nuclear power projects, pharmaceutical plants, oil and gas facilities and rail infrastructure, particular licences and certificates may be required. If there is any doubt ask the design team for their advice, then manage the process.

Establish site

Once the design has been finalised and the contracts signed, the project is ready to go on site. It is imperative that the site set-up process is carried out and completed in the most efficient manner prior to the start of the main construction works. The issues that the project manager must be aware of and monitor with the contractor at this stage are not only practical and physical operations but also administrative plans and procedures agreed by the parties. The areas where the project manager is to agree and monitor site set-up are:

■ Site boundaries clearly identified with the contractor.

■ Establish the contractor's proposal for security.

■ Establish the contractor's proposal for emergency plans in case of fire or any serious incidents.

■ Establish the contractor's proposal of site accommodation; specifically the suitability of welfare facilities.

■ Carry out a survey of existing conditions of the site and the adjacent properties. Record any relevant issues.

■ Establish with the contractor the administrative procedures such as request for information (RFIs), confirmation of verbal instructions (CVIs), daily returns, daily diaries, faxes, e-mail facility, drawing issues, etc. This activity is one of the most important action to do as it will set out the communication route between all the participants throughout the project. The findings and agreements with the contractor should be recorded by the project manager and distributed to all professionals involved.

■ Ensure that the contractor is aware of, and is attending to, any issues that may be present due to neighbours being close to the site including the terms of any party wall awards or rights of light issues.

■ Ensure that the contractor has clearly identified the health and safety risks that exist on the site.

■ Ensure all signage is displayed correctly.

The above issues are to be agreed with the contractor. The project manager cannot dictate how the contractor is to set up the site. The project manager's role must be advisory and thus monitor that the correct actions as agreed are being implemented.

Control and monitoring systems

It is the project manager's prime duty to make sure that all necessary control and monitoring systems are properly set up and implemented by the contractor.

The project manager should endeavour that these systems produce the most appropriate information and reports, on a regular and timely basis, so that they can be used to monitor and manage the project to its successful conclusion.

By carrying audits and checks of the systems, the project manager must be fully satisfied with the accuracy of the information produced and that it does indeed indicate the 'real' position at any point in time and, where appropriate, accurately forecast the final position for the project.

Contractor control and management systems will generally be (but not limited to) the following:

- quality management system

- schedule management system

- quality control system

- cost monitoring and management system

- health, safety and welfare system

- environmental management system including waste management

- ICT system

- document management system.

It is of absolute importance that the project manager fully understands the relevance of the information being produced from these systems. The project manager must proactively use this information to manage the contractor and the project team through the regular management meetings. The aim is not only to understand where the project is and where it is ultimately going, but also to identify any potential problem areas at such a sufficiently early stage so that any rectification procedures and/or mitigation measures can be taken to ensure best delivery of the project.

Contractor's working schedule

The project manager has a duty to the client to monitor the performance of the contractor. In order to adequately carry this out, the project manager needs to ensure the contractor has prepared a construction schedule (working schedule) in sufficient detail to enable the construction works to be closely monitored.

The project manager needs to receive and review the contractor's working schedule prior to the commencement of the works in order to:

- check it complies with the client's time requirements

- check it acknowledges any restraints imposed on the construction of the works

- ensure that the level of detail is appropriate for the illustrating the progress of the works

- ensure it suitable for monitoring the progress of the works

- confirm the sequencing and logic of the schedule.

The working schedule must incorporate an information requirement schedule so that it realistically informs the project manager when outstanding design information is required in order for the contractor to achieve the schedule dates. Regular reports recording the progress achieved against the schedule must be received from the contractor, and a progress status agreed with the contractor.

Any rescheduling of the works necessary to recover delay situations need to be received, reviewed and agreed. In addition to the detailed analysis of progress, the project manager should examine high-level progress trends to obtain an overall view of project status. This can involve graphically comparing accumulative planned progress against actual achieved. Usually the contract requires that the contractor prepare a contract schedule that becomes part of the contract. This schedule does not normally go into a great deal of detail, as timescales, dependencies and interfaces have yet to be agreed with subcontractors to the main contractor. The project manager will need to obtain a working schedule detailing all specific sections of work. Rescheduling required as a result of changes or slippages will be required to show how time can be recovered or the effect on the completion dates.

It is the duty of the project manager to not only to monitor the contractor's progress, but also to monitor any work being undertaken by other advisors, suppliers or companies that have an independent input into the completion of the project. These should all be monitored against the development schedule with its own milestones and targets. The project manager is managing the overall project for the client and its successful delivery.

Value engineering (related to construction methods)

Value engineering (VE) is an exercise that most of the project team undertakes as the project develops, by selecting the most cost-effective solution. However, VE is about taking a wider view and looking at the selection of materials, plant, equipment and processes to see if a more cost-effective solution exists that will achieve the same project objectives.

VE should start at project inception where benefits can be greatest, however the contractor may have significant contributions to be made as long as the changes required to the contract do not affect the timescales, completion dates and incur additional costs that outweigh the savings on offer. There is, however, still a place for VE, especially at the start of construction. The application of the job plan (see Table 4.2) remains consistent, but the detail available is obviously more than during the design and pre-design stages. The 'results accelerators' still act as useful guides to VE at the construction stage (see Table 4.3). In all of this it is most important to remember the relationship between cost and value: value is function divided by cost. Concentrating on the function of the project or product will avoid mere cost cutting.

Table 4.2 Value engineering job plan

Information
Function analysis
Speculation
Evaluation
Development
Recommendation
Implementation

Table 4.3 Result accelerators

Avoid generalities
Get all available costs
Use information from best source
Blast, create and refine
Be creative
Identify and overcome roadblocks
Use industry experts
Price key tolerances
Use standard products
Use (and pay for) specialist advice
Use specialist processes

The project manager must take a proactive role in both giving direction and leadership in the VE process, but must above all ensure that time and effort is not wasted and does not have a detrimental effect on the progress of the project. An example of a VM framework has been included in Appendix 10.

Management of the supply chain

The contractor has overall responsibility for the management of the supply chain to meet the contractual obligations. The project manager has the duty to ensure that this chain is being effectively managed so as to avoid any potential delay, unnecessary cost implications or any other adverse effect on the delivery of the project. This is an important issue as it is so very often the case that problems further down the contractual chain can be responsible for long delays and/or major disputes right back up through the whole chain. They can potentially result in the deterioration of relationships with the contractor and have a knock-on effect with not only the contractor's performance, but also the performance of the project team as a whole.

Duties and responsibilities should include:

- Receive and understand the details of the contractor's supply chain and the controls to manage it.

- Establish key members and linkages within the chain.

- Receive and interrogate reports from the contractor of the ongoing progress, including any reports from procurement managers and expeditors.

- Implement a regular monitoring system to check the progress of key suppliers or subcontractors (against the contractor's delivery schedule) so that there are timely warning signals of any potential delays or failures that could have an adverse effect on the progress and financial stability of the project.

- Agree with the contractor any appropriate remedial action that may be needed to rectify any problem areas.

Risk register

The risk register (see Appendix 9) is a document that should be prepared at the earliest stages of the project, identifying potential risks throughout the project. This register should be reviewed and updated according to circumstances, and stages of the contract. At the construction stage the risk register should be reviewed to include any new construction risks.

In addition to monitoring those construction-related risks previously identified in the project-wide risk register, the project manager should require the contractor to instigate and maintain a risk management system for those risks likely to impact on the actual construction works. The project manager should also require the contractor to:

- establish a fully detailed listing of construction risks

- determine the likely probability and impact of each risk

- review the risks with the project team

- prepare method statements and action plans demonstrating how risks will be mitigated or managed out

- identify, and inform, the person responsible for managing each risk

- prepare contingency plans for any key risks having a significant impact

- regularly review and report on the status of risks.

Benchmarking

In certain circumstances, particularly when framework or partnering agreements are in place, it may be appropriate to employ benchmarking of a contractor's performance against the best industry practice. A major difficulty of benchmarking in construction is locating base data that allows for meaningful comparisons. Since 1998, as part of the annual production of statistical information gathered from contractors, the British government collected measures of key performance indicators. These provide the widest sourced comparators currently available to benchmark individual companies against the industry average levels of performance.

A number of construction clients commission their own research to assemble from other similar organisations meaningful performance data that allows them to carry out benchmarking of the companies they use.

Benchmarking is closely associated with the concept of continuous improvement, and a company's performance can be monitored over time to confirm that measures introduced to lead to improvement are effective.

Change and variation control

The project manager should carry out the following tasks to control variations:

- Monitoring and controlling variations which result from changes to the project brief to be avoided whenever possible (Figure 4.4) or design/schedule modification (e.g. client's request, architect's or site instructions) must follow a procedure which:

  ○ identifies all consequences of the variation involved

  ○ takes account of the relevant contractual provisions

  ○ defines a cost limit, above which the client must be consulted and, similarly, when specifications or completion dates are affected

  ○ authorises all variations only through the appropriate change procedure.

- Identifying, in consultation with the project team, actual or potential problems and providing solutions which are within the time and cost limits and do not compromise the client's requirements, with whom solutions are discussed and approval obtained.

- Checking the receipt of scheduled and/or ad hoc reports, information and progress data from project team members.

The main effect on the reduction of claims or variations is to ensure the brief is clearly defined, and the contract documents and drawings accurately and completely reflect the detail.

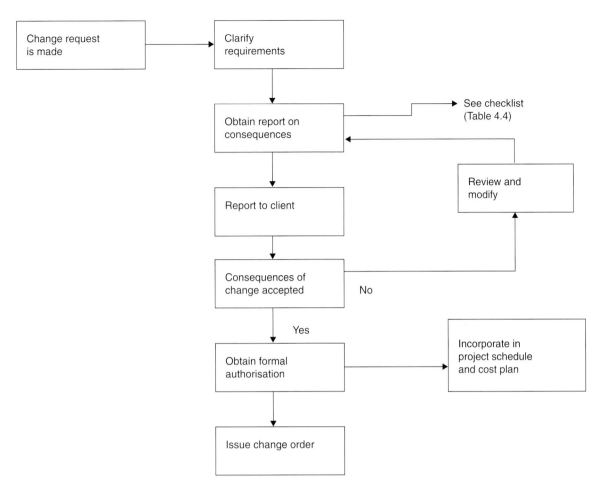

Figure 4.4 Changes in the client's brief.

Table 4.4 Changes in the client's brief: checklist

	Activity	Action by
1.	Request for change received from client	Project manager
2.	Client's need clarified and documented	Project manager
3.	Details conveyed to project team	Project manager
4.	Review of technical and health and safety implications	Consultants and project manager
5.	Assessment of programme implications	Planning support staff and project manager ·
6.	Evaluation/calculation of cost implications	Quantity surveyor
7.	Review engineering services commissioning	Commissioning manager
8.	Preparation of report on effect of change	Project manager in consultation with consultants
9.	Reporting to client	Project manager
10.	Consequences accepted/not accepted by client	Project manager
11.	Non-acceptance: further review/considerations as per items 4–6 and action items 7 and 8	Project manager assisted by consultants
12.	Further reporting to and negotiation of final outcome with client	Project manager assisted by consultants
13.	Agreement reached and formal authorisation obtained	Project manager
14.	Incorporation into project programme and cost plan (budget)	Project manager and quantity surveyor
15.	Change order issued (see Appendix 16)	Project manager and client

Managing change control at design development stage is far more effective than managing the process when construction is in progress. Circumstance-driven changes, mistakes or unknowns have to be effectively managed on the basis that in many instances time is more expensive than the material change. Some form of authorisation needs to be agreed (financial limits) so that instructions can be given without having to refer every change back to the customer for approval.

The project manager will need to maintain a register of changes and variations, cross-referenced to contractor's RFI notices, and possibly contract claims. The register should include budget costs and final costs for reporting to the client on a regular basis.

Accurate, detailed daily diaries will need to be kept, complete with plant, labour and material deliveries so that consequential costs can be identified.

In dealing with the effects and costs of variations, the project manager will need, where possible, to agree costs before issuing an instruction. It is also wise to agree, again where possible, that work will be undertaken with no overall effect on the schedule. It is vital to carefully recording events and situation at the time.

Procedurally, the project manager must inform both the design consultants and the main contractor that all variation instructions must be in the correct written form and *must* only be issued via the project manager unless the project manager is the appointed contract administrator to the main contractor. To avoid unnecessary complications in agreeing valuations and accounts, it is imperative that the variation instructions are issue from one source. Design consultants must raise (in writing) RFI to the project manager who will in turn issue the instructions to the contractor. *All* variations *must* have an instruction (in writing) against them in order to be valued.

5 Construction stage

Interlinking with previous stages

The change from pre-construction to the construction stage is a watershed. It is the culmination of all the pre-construction effort that allows the actual work to commence on the site. With this change, the duties of the project manager also change and this section sets out the project manager's tasks on the premise that the project manager then supervises the construction and final delivery of the project.

The move to construction must be managed as a seamless operation and must recognise and enact any key policy or strategy decisions that will have been taken during the earlier stages of the project cycle. Decisions will have been taken in such areas as the client's main requirements, planning requirements, whole life-cycle constraints, value engineering, procurement methods, early contractor or specialised subcontractor/supplier involvement, health, safety and welfare, environmental issues, and so on. The procedures and responsibilities for all these have been dealt with in the earlier sections of this *Code of Practice* and now must be effectively implemented during this dynamic stage of the project.

Ideally, construction will be undertaken in accordance with a predetermined and detailed construction schedule. This efficient ideal is possible on many projects, especially those working with well-developed designs. However, some projects, especially those using innovative technologies and designs, benefit from further initiatives and development of the project. In these cases a proactive input is needed from the project manager and all the members of the project team to search out and consider further practical ideas that will enhance the end product.

It is the project manager's overriding role during the construction phase to provide the team with the necessary strong and proactive leadership. The project manager has to steer the project to completion through continuous measurement of performance against time, quality and costs and to carry out all necessary actions to ensure the team's successful delivery of a project, that not only meets the client's satisfaction but that also exceeds expectations.

Responsibilities of project manager at this stage

To be an effective leader

The project manager needs to demonstrate soft skills as well as hard skills:

- Hard skills generally include planning, scheduling, organisational ability, report writing, information assembly, cost control, innovation, decision-making and prioritisation.

■ Soft skills include leadership, motivation, communication, interpersonal skills, personality, team-building abilities, honesty, integrity, and sense of humour. In context of the growing acceptance of the criticality of these soft skills, further guidance on elements such as leadership, motivation and team building have been provided in Appendix 21.

To set the project objectives

The project manager has the responsibility to define the primary objectives for the project. From these the project manager has to develop individual objectives, team objectives and the project's general objectives in order to achieve the primary objectives. The project manager has then to be able to effectively communicate with the team members and obtain their commitment to achieve them. These should include:

■ meeting the client's objectives within the contract, i.e. not at all costs

■ fair treatment to all parties to the project

■ customer focus

■ ensuring a delighted client.

To ensure achievement of objectives

The project manager must clearly adhere to the project success criteria. The project manager must maintain the measurement against the progress and then pro-actively manage the project to ensure success.

To achieve client satisfaction

This has to be the prime responsibility of the project manager.

Role of project team members

Although the precise contractual obligations of the project participants vary with the procurement option adopted, the project participants must carry out certain essential fundamental functions.

Client

Traditionally, the client has had a relatively nominal direct involvement in the construction works, however as more and more client project teams are being constituted with extensive construction background, the role of the project manager in managing the client's expectations is also expanding. There is now a greater emphasis on the client having more involvement during the construction stage with their primary interests being:

■ Ensuring that the build quality is acceptable, taking advice from the project manager where appropriate.

■ Progress of the works is to schedule and in a logical fashion.

■ Understanding potential effects of client changes to the construction stage progress.

■ Managing internal stakeholders in terms of decision making to help with the progress on the site.

- Ensuring security, environmental friendly and safe working practices are adopted.

- Satisfying themselves that the contractor's performance is in accordance with the contract.

- Making sure the obligation to pay all monies certified for payments to consultants and the contractor(s) is being carried out.

Project manager

The project manager has a role which is principally that of monitoring the performance of the main contractor and the progress of the works, and involves the following activities (some of which may have been accomplished in the pre-construction stage):

- Ensuring contract documents are prepared and issued to the contractor.

- Ensuring the contracts are signed.

- Arranging the handover of the site from the client to the contractor.

- Reviewing the contractor's construction schedule and method statements.

- Ensuring the contractor's resources are adequate and suitable.

- Ensuring procedures are in place and being followed.

- Ensuring site meetings are held and documented.

- Monitoring construction cash flow.

- Reviewing progress with the contractor.

- Monitoring performance of the contractor.

- Ensuring that the construction phase health and safety file is being maintained.

- Ensuring design information required by contractor is supplied by consultants.

- Establishing control systems for environmental sustainability, time, cost and quality.

- Ensuring site inspections are taking place.

- Confirm insurance cover on the works.

- Managing project cost plan.

- Ensuring that the client meets contractual obligations (i.e. payments).

- Reporting to client.

- Managing introduction of changes.

- Ensuring statutory approvals are being obtained.

- Ensuring all relevant legal documents are in place (such as collateral warranties and performance bonds among others).

- Review construction risks.

- Establish mechanisms for dealing with any claims.

- Monitor for potential problems and resolve before they develop.

Part 1 Project management

Design team

The design consultants are responsible for the following:

- Providing production information (i.e. details of building components).

- Commenting and approving working drawings being provided by specialist contractors.

- Responding to site queries raised by the contractor.

- Inspecting the works to check compliance with the drawings and specification.

- Inspecting the works to check an acceptable quality standard has been achieved.

Most building contracts refer to a contract administrator, usually the design team leader or the project manager, who is the formal point of contact between the project team and the contractor, and who has a contractual obligation in relation to the issuing of formal instructions to the contractor; these include the following:

- issuing of design information

- issuing of variations

- instructions on standards of work and working methods

- arbitrating on contractual issues

- issuing practical completion certificate.

Quantity surveyor

The quantity surveyor has a duty to:

- measure the value of work executed by the main contractor

- agree monthly valuations with the main contractor

- agree the final account with the main contractor.

The quantity surveyor has a separate responsibility to the client, usually through the project manager, for reporting on the overall financial aspects of the project.

Contractor

The contractor has several statutory and contractual responsibilities that must be enacted in order to allow the construction of the project to proceed. Depending on the precise form of contract, these responsibilities will vary but will generally include the following:

- executing the contract agreement between the employer and contractor

- submitting the requisite health and safety documentation

- actioning compliance with the requirements of the CDM Regulations of 2007 (see Appendix 2)

- implementing the site waste management plan as required under the Site Waste Management Plans Regulations of 2008 (see Appendix 27)

- producing documentary evidence of all insurance policies as required by the contract

- enacting all parent company guarantees, bonds, warranties, indemnities and third party rights as required by the contract

- actioning any statutory notices and consents such as planning requirements, hoarding licences, scaffold licence

- actioning any third party notices, licences and consents such as tower crane over-sailing agreements

- gaining any necessary consents from the employer such as subletting any part of the works

- providing the programme of works with all relevant method statements and activity schedules

- mobilising all necessary labour, subcontractors, materials, equipment and plant in order to commence the construction works in accordance with the contract.

Construction manager

A client may decide on a construction management route, directly employing a construction manager as a consultant acting as an agent with expertise in the procurement and supervision of construction and not a principal. In this arrangement the construction manager's role is the following:

- To determine how the construction works should best be split into packages

- to produce detailed construction schedules

- to determine when packages need to be procured

- to manage the procurement process

- to manage the overall site facilities (such as access, storage, welfare, etc.)

- supervise and co-ordinate the works package contractor's execution of the works.

Management contractor

In the managing contracting arrangement, a management contractor acting as a principal would have the additional direct contractual responsibility for the performance of the works package contractors.

Subcontractors and suppliers

Subcontractors have specialist expertise, usually trade related (i.e. mechanical or electrical installations, lift installation, joinery and demolition), for the supply and installation of an element of the total works.

Subcontractors may be either nominated or named by the consultants or selected and appointed directly by the main contractor, known as domestic subcontractors. If nominated, the client carries some risk in respect of the subcontractor's performance.

Suppliers provide certain materials, components or equipment for others to install.

Labour-only subcontractors provide only labour to carry out the installation of materials, components or equipment provided by the main contractor (i.e. carpenters, bricklayers and plasterers).

Due to their specialist knowledge, subcontractors have an increasing design responsibility for the technical design related to their installations (may include fixing details, fabrication details, co-ordination with other installations).

There is a general obligation on all the project team to ensure the site is safe, although legally this falls to the principal contractor under the CDM Regulations.

Other parties

A large number of other bodies will be involved during the course of the construction works, these include the following:

■ building control officer

■ highways authority

■ environmental health officer

■ fire officer

■ Health and Safety Executive

■ planning officers

■ archaeologists

■ trade unions

■ landlord's representatives

■ funder's representatives

■ police.

Team building

Traditionally, it has always been easy to execute contracts and projects using the contract and the specific duties and responsibilities for each party. However, this rigid approach has more often than not brought about an adversarial environment between usually the contractor and the client's design team.

Construction is a people business and thus communication is the key to a successful project. This involves the project manager heading up the professional design and construction team (including the contractor) and building trust between all the parties. The project manager will be responsible for the ultimate outcome of the project and thus it is in his interest that he has a united team working towards the same goal. There are numerous individual methods of teambuilding which can be adopted by referring to numerous literature published on management. However, the most effective and the critical period for the project manager to drive home his stance on the team is at the outset of the project concept, as and when designers and consultants together with the contractor are brought on the team. Further guidance on this issue is provided in Appendix 21.

Regular progress meetings and workshops (both formal and informal) can assist in developing the bond between all parties. Importantly, during the construction stage, the team must have a hands on (immediate) approach in resolving, assisting and thus eliminating any issues hindering the smooth process of construction. The construction business is not about individuals, but teams both at pre-construction stage and also during construction.

Health, safety and welfare

The project manager must be aware and be able to manage the processes and requirements associated with health, safety and welfare legislative requirements. The four main individual or groups deemed to be holding key responsibilities under the CDM Regulations (see Appendix 2 for further details) are:

- the client

- the designers

- the CDM co-ordinator

- the principal contractor.

The project manager has a responsibility to be monitoring the activities and actions being executed by the above four entities. This does not mean that the project manager will be held liable for any wrongdoing, but simply the management of the project from the inception stage through design, construction and finally occupation, must have a health and safety aspect within the strategy, design and delivery process, eliminating risks at each stage. There are number of tools for monitoring the successful process of health and safety of individual potentially 'risky' design solutions and on-site construction method issues. The main tools are risk assessments, risk workshops, method statement analyses and the health and safety file (pre-construction, during construction and post-construction).

During the pre-construction stage the responsibility of health and safety is mainly on the design team co-ordinated by the CDM co-ordinator (where applicable). The management process must be considered by the project manager. During construction, the principal contractor is responsible for safety and welfare on the site. The principal contractor is responsible for preparing a construction-phase health and safety plan. This should be updated regularly. At completion of the project, the CDM co-ordinator prepares a concise health and safety file on the built product highlighting any potential risks to the end user. This document is prepared on the basis of the information provided by the principal contractor.

Health, safety and welfare is the collective responsibility of all individuals in the construction business. The project manager must take an active role in monitoring the process. The project manager must emphasise the importance of health and safety considerations to the client, most importantly, and also to the design and construction team. See Appendix 2 for further information on health and safety including the CDM Regulations.

Environmental management systems

Environmental statements

Environmental concerns will increasingly affect our projects. This is especially the case with the pressure to develop brownfield sites and reuse old sites. The cost of addressing contaminants or other environmental issues can add significant costs and increase the duration of project. Planning authorities are also more likely to instruct environment studies and restraints as part of the planning process, all of which must be incorporated into the project during the construction stage. It is the project manager that has overall responsibility to ensure compliance with these aims, objectives and constraints. The project manager will need to:

- Understand and act on the environmental impact assessment; see Appendix 11.

- Ensure proper environmental advice is available.

- Ensure that the contractor is complying with the environmental statement; see Appendix 11.

- Seek and ensure action by the contractor of any remedial actions should they be necessary to comply with environmental considerations.

Contractor's environmental management systems

The contractor must establish his own environmental management systems (EMS), but it is for the project manager to ensure that it is being managed properly and is progressing sufficiently to achieve all EMS objectives. Therefore the project manager should:

- Receive details of the contractor's EMS and the environmental plan (EP) specific to the project.

- Ensure that the contractor has set up all necessary procedures and structure to manage the EMS and implement the objectives of the EP.

- Check that the contractor's environment management plan matches the aims and objectives of the environmental statement.

- Agree with the contractor any further aims, specific targets or initiatives that will maximise sustainability of the project and minimise the detrimental impact of the construction process.

- Proactively monitor the progress of the contractor to maintain his proposals and objectives.

Compliance with site waste management plans 2008

In April 2008 site waste management plans (SWMPs) became a legal requirement for all construction and demolition projects in England, for projects valued over £300,000. A SWMP provides a framework for managing the disposal of waste throughout the life of a construction project. In essence, it should contain the following information:

- ownership of the document

- information about who will be removing the waste

- the types of waste to be removed

- details of the site(s) where the waste is being taken

- a post-completion statement confirming that the SWMP was monitored and updated on a regular basis

- an explanation of any deviation from the plan.

Generally, the SWMP will be instigated by the client at the pre-construction stage, where the designers will also have to provide the required information. At the construction stage the document becomes the responsibility of the principal contractor. Further information on SWMPs is provided in Appendix 27.

Inspection/monitoring of the works

Once the project is underway on the site, regular inspections and monitoring of progress is to be carried out by the project manager. There is a fine line as to how involved the project manager should become with the everyday issues facing the contractor, and thus the relationship, as mentioned previously, will determine the appropriate approach.

It is the project manager's responsibility to arrange from the outset progress meetings at regular intervals. During these meetings the contractor will present a report as to progress on the site with any relevant design issues which will require resolving. If necessary, separate design meetings should also be set up. The reporting process to the project manager must not be restricted to the contractor but also to all designers and consultants. It is at these forums that the project manager must manage and ensure all parties are working together and achieving individual target dates for producing information and maintaining progress against the schedule.

Notwithstanding formal progress meetings, the project manager should also visit site regularly and spend limited time at the site discussing progress with site staff and chasing up the appropriate individuals for information and progress.

Reporting

A fundamental aspect of the project management role is the regular reporting of the current status of the project to the client. The project manager needs to ensure an adequate reporting structure and calendar is in place with the consultants and contractors. Frequency and dates of project meetings need to be co-ordinated with the reporting structure. Reporting is required for a number of reasons:

■　to keep the client informed of the project status

■　to confirm that the necessary management controls are being operated by the project team

■　to provide a discipline and structure for the team

■　as a communication mechanism for keeping the whole team up to date

■　to provide an auditable trail of actions and decisions.

Progress reporting should record the status of the project at a particular date against what the position should have been; it should cover all aspects of the project, identify problems and decisions taken or required, and predict the outcome of the project. The project manager needs to receive individual reports from the consultants and contractor and summarise them for the report to the client. The detailed reports should be appended as a record. Typical contents of a project manager's project report would contain the following:

■　an executive summary

■　legal agreements

■　design status

■　planning/Building Regulation status

■　procurement status

■　construction status

■　statutory consents and approvals

■　project schedule and progress

■　project financial report

■　　variation register update

■　　major decisions and approvals required.

Trends shown visually are an excellent mode of conveying information to clients and senior management.

Public liaison and profile

The client would probably have set out his overall public relations and liaison strategy during the pre-construction stages of the project. In reflection to this, the project manager should take a leading role in 'local' public relations during the construction stage. This will improve the public's perception of the construction industry in general. Such activities or actions should include the following:

■　　ensuring that there is no local nuisance or negative impact arising from the project

■　　maintaining good housekeeping both on-site and in the immediate off-site area

■　　erecting informative scheme boards and public viewing platforms

■　　ensuring that the contractor takes part in a local or national 'considerate contractor' scheme

■　　taking awareness initiatives with local schools

■　　attending local public meetings to raise the profile of the project

■　　organising site visits for local schools, residents and business people

■　　partaking in local environmental schemes or issues

■　　being involved with fundraising for local charities or causes.

As discussed previously, a pre-agreed 'communication plan; will provide clear directions in this context (see Appendix 25).

Commissioning and operation and maintenance manuals

Commissioning

The main commissioning and putting to work issues are covered in Chapters 6 and 7. This section covers the construction stages.

The project manager should receive the contractor's commissioning schedule within the early stages of construction in order to satisfy himself that it is properly co-ordinated with the building works schedule. (As an example, the balancing of the heating and air conditioning system can only take place when the building envelope and internal spaces have been secured.) A problem may occur in that in many instances the building services contractor is a subcontractor to the main contractor and that the subcontract may not be in place at this early stage of construction. In this case, the main contractor will need to identify the logic and sequence of the commissioning.

Operating and maintenance manuals

The CDM Regulations now cover the operating and maintenance manuals, and it is the CDM co-ordinator's role to ensure that they are delivered as part of the health and safety file. These manuals should also include details of the complete building with input from all of the design team. The project manager has to monitor the progress that the CDM co-ordinator is making on assembling these files and, if needs be, ensure that all necessary actions are taken to expedite their completion with active co-operation from the contractor. It is a legal responsibility of the contractor to co-operate with the CDM co-ordinator and to comply with any reasonable request from the CDM co-ordinator in order to enable completion of the health and safety file including the operating and maintenance manuals.

Payment

A vital part of the construction process is to ensure that the contractor and subcontractors receive regular payments for their work. The project manager has a role in attempting to avoid disruption caused by contractors failing on the project. There are a number of actions the project manager can take:

- checking of the financial standing of contractors prior to their appointment
- ongoing monitoring of the financial position of contractors
- ensuring all payments due are paid promptly by the client.

Generally, the contractor makes a monthly application for payment. The quantity surveyor values the work, the contract administrator certifies it and the client pays for it within a stipulated period of the application/certification date. The project manager has an important role in ensuring that the client honours obligations to pay contractors against certificates authorised by the contract administrator.

As projects become larger and more complex, so do the means of finance. These include the following:

- public–private partnership (PPP)
- private finance initiative (PFI)
- design built finance operate (DBFO)
- build own operate transfer (BOOT)
- cost plus, reimbursable, target cost, cost plus fee.

These forms of contract (and variations thereof, see Appendix 28) are more likely to have their own 'tailored' methods and formats for payments to the contractor or concessionaire, but set out below are the more common methods of payment for traditional or design and build contracts:

- **Valuations.** The traditional method of payment has been a physical measuring of the works carried out on site and the quantity of work costed against the rates in the bill of quantities. Carried out jointly by the main contractor and quantity surveyor. This is usually done monthly. The contract administrator issues an interim certificate for the amount due to the client and the client has to make payment to the main contractor within the period stated in the contract.

- **Milestone.** Tenderers, as part of their tenders, are asked to break down their total price into a number of sums against predetermined milestones. Milestones usually being the completion of elements of the construction works (e.g. completion of the structure up to a certain level). There are normally likely to be 20–40 milestones. Acknowledgement that a milestone has been achieved by the contract administrator will release payment of the sum to the contractor. This can sometimes be called an 'activity schedule' (NEC contract).

- **Stage.** Similar to the milestone payment, but there are likely to be far fewer stages (e.g. completion of superstructure, achievement of a watertight building).

- **Earned value.** Regular payments made in accordance with an earned value system. Payment will be related to the actual progress position achieved on the works. As the value of payments is based on the schedule assessment of progress it avoids the need to separately carry out monthly measurements of works carried out.

- ***Ex-gratia.*** Although not a formally recognised method of payment, in certain extreme cases when lack of cash is preventing a contractor from carrying out his obligations under the contract, a special one-off payment may be made with the client's agreement. This is an *ex-gratia* payment made in advance of the normal payment procedure to ensure certain works are carried on order to recover or prevent a delay situation or to expedite certain materials. In cases this might be accompanied by a pre-payment insurance bond. It is important that if payment is for materials or equipment that ownership is clearly established in case of insolvency of the contractor.

6 Engineering services testing and commissioning stage

Client's objectives

At this stage the project team should prove that the engineering installation has been installed correctly, in a safe manner, and that it performs to the requirements of the design. The project manager's objective is to ensure that the commissioning of the separate systems is properly planned and executed, so that the installation as a whole is fully operational at handover without delay to the programme and that any fine-tuning necessary after handover is carried out in liaison with the client and/or user.

Interlinking with construction

It must be stressed that the location of this chapter does not mean that the activities involved only take place at the end of the construction stage. Engineering commissioning is a very important part of the construction process and must be addressed and considered very early on within the project. The following are suggested activities that must be considered well before this stage:

- Decide the most appropriate time within the project to appoint the commissioning contractor and his role/scope of work.

- Where appropriate, appoint a commissioning contractor to review the design drawings and working drawings to ensure commissionability.

- Ensure consultants clearly identify testing and commissioning requirements.

- Ensure consultants/client identify performance/environmental testing requirements

- Ensure that the project schedule includes sufficient time to undertake the specified commissioning and, in particular, the additional time required for any performance/environmental testing and statutory testing to authorities.

- Clearly identify the method of presenting, recording and electronically storing 'as-installed' information.

- Although not strictly part of engineering commissioning, ensure that the requirement for specialist maintenance contracts for equipment is carefully considered prior to awarding tenders for such equipment.

Commissioning generally

Commissioning is carried out in four or sometimes five distinct parts:

- static testing of engineering services

- dynamic testing of engineering services

- performance testing of engineering services (not always undertaken)

- undertaking statutory tests for various authorities

- client commissioning.

Note that performance testing also includes environmental testing. The first four items, engineering services testing, commissioning, performance testing and statutory tests are part of the construction design and installation phases of the project. Client commissioning is an activity predominantly carried out by the client's personnel assisted, where required, by the consultants. This is dealt with in Chapter 7. The engineering services testing and commissioning process objectives and main tasks are as described within this chapter.

Procurement of commissioning services

Smaller projects

There are many ways to procure the commissioning specialist. On smaller projects, via the main contractor, the mechanical and electrical subcontractors are most likely to be responsible for the testing and commissioning of their installations. Electrical contractors will normally use their own resources, except where specialist items of equipment require the manufacturer to assist with their testing. Mechanical contractors will usually appoint a commissioning specialist to work on their behalf. Again, where specialist items of equipment are installed, the mechanical contractor will request the manufacturer to assist with testing where appropriate. However, it should be noted that often these commissioning specialists are no more than balancing engineers. This is fine for simple installations, but where more complicated systems are involved or specific commissioning and performance tests are required, their management and execution may not be adequate. Careful specification of the requirements within the design documentation is required when tendering the installation work. This is all too often ignored or given insufficient time and effort which inevitably creates problems later in the construction process.

Larger projects

On larger projects, the method of procuring the commissioning specialist can take many forms. In traditional forms of contract it can again be via the main contractor/services contractor, however, in construction management or similar forms of contract, a specialist commissioning contractor is often appointed. This commissioning contractor normally fulfils one of two roles: the role of managing the testing and commissioning process (the actual work being done by the installation contractors as detailed for small projects above), or the role of undertaking the commissioning work. In this latter role, the point of delineation for testing/commissioning between the installation contractor and the commissioning contractor is usually at the end of static testing and the start of dynamic testing. See below for a definition of these terms. This latter role is gaining in popularity for the following reasons:

- It provides a degree of independence to the commissioning process.

- The commissioning contractor is under the control of the construction manager/managing contractor and reports directly to them, giving greater control and transparency to the process.

In either role, the benefit to the project is that the commissioning contractor can be brought into the project very early to manage the whole testing and commissioning process.

Role of the commissioning contractor

Below are some of the activities that can be included within the scope of work for the commissioning contractor:

- Review the design drawings near the end of design to ensure familiarity with the design intent and to add their expertise in to the commissionability of a scheme.

- Ensure that the testing and commissioning is correctly specified in the tender documentation.

- Review the services contractor's working drawings for commissionability.

- Set up the testing and commissioning documentation to create consistency between the various contractors.

- Define the method, media type, style and content of the as-installed information to create consistency between the various contractors.

- Manage the specialist equipment manufacturers' tests.

- Liaise with building control and other organisations to witness relevant statutory tests (including insurer's tests).

All of these functions are often given insufficient thought on projects, so if they are not to form part of the commissioning contractor's brief, then it should be recognised that some other part of the project team should undertake this work.

The testing and commissioning process and its programming

Flowcharts relating to the various stages of testing, commissioning and performance testing are given in Figures 6.1 and 6.2 below. It is important for the project manager to understand the differences between the terms testing, commissioning and performance testing, and for him to ensure that the programme has sufficient time within it to enable these activities to be undertaken. Unfortunately, with this stage of the project being so close to handover, there is often pressure to gain time by shortening the testing, commissioning and performance/environmental testing programme. This should be strongly resisted. Rarely, if ever, after the project will such an opportunity exist to fully test the services to ensure that they work individually, as a system, and, that they work under part-load and full-load conditions. Many problems with respect to the underperformance of services within an occupied building can be related back to either insufficient quality in the testing and commissioning, or, insufficient time to test and commission.

It should also be borne in mind that various statutory services will need to be demonstrated to building control (or the relevant government department if a Crown building) and insurers. Time should be allowed for within the programme since these activities are often taken as separate tests after the main commissioning has been undertaken.

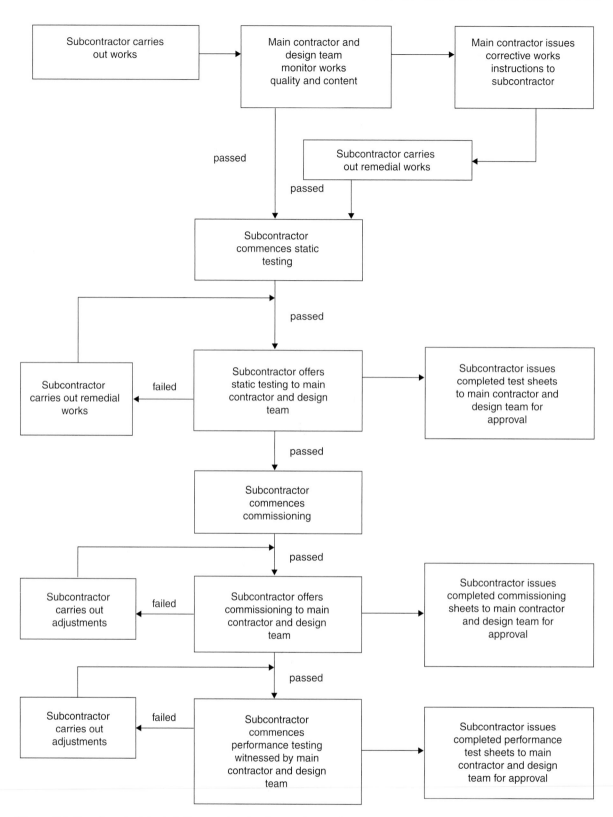

Figure 6.1 Small project installation works checks, testing and commissioning process and sign off.

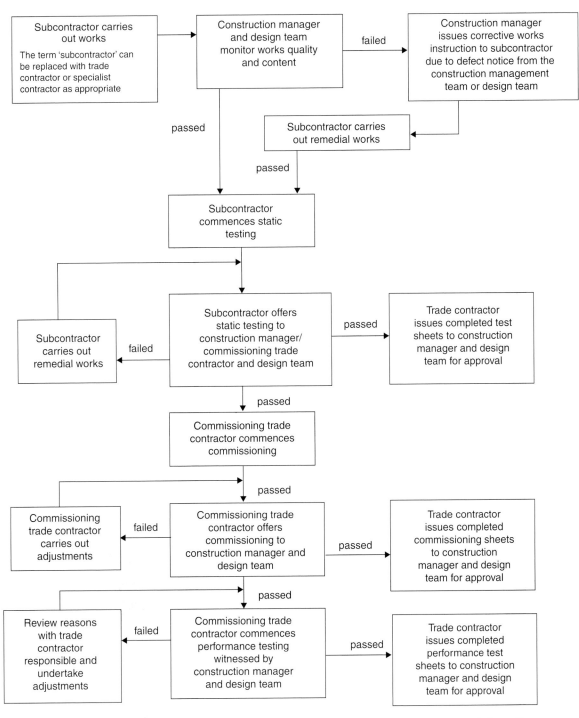

Figure 6.2 Large project installation works checks, testing and commissioning process and sign off.

Differences between testing and commissioning

Testing

During the services installation various testing will be undertaken known as 'static testing'. This testing is normally undertaken to prove the quality and workmanship of the installation. Such work is undertaken before a certificate is issued to 'liven' services whether electrically or otherwise. Examples of systems tested are:

■ pressure testing ductwork and pipework

■ undertaking resistance checks on cabling.

Commissioning

On completion of the static testing, dynamic testing commences: this being the commissioning. Commissioning is undertaken to prove that the systems operate and perform to the intended design and specification. This work is extensive and normally commences by issuing a certificate permitting the installation to be made 'live', i.e. electrical power on. After initial tests of phase rotation on the electrical installation and checking fan/pump rotation (in the correct direction), the more recognised commissioning activities of balancing, volume testing, load bank testing and so on begin.

Performance testing

On completion of the commissioning, performance testing can commence. Some may not distinguish between commissioning and performance testing. However, for programming purposes it is worth distinguishing the difference between commissioning plant as individual systems and undertaking tests of all plant systems together, known as performance testing (and including environmental testing). Sometimes this performance testing is undertaken once the client has occupied the facility, e.g. for the first year because systems depend on different weather conditions. In such cases, arrangements for contractor access after handover to fine-tune the services in response to changing demands must be made. However, for some facilities it is desirable, if not essential, to simulate the various conditions expected to prove that the plant systems and controls operate prior to handover, e.g. in the case of computer server rooms.

Main tasks to be undertaken

To assist the project manager, the following has been provided to summarise the main tasks to be carried out during the three main stages of pre-construction, construction and post-construction.

Pre-construction

The following items will need to be confirmed:

- The consultants/client recognise engineering services commissioning as a distinct phase in the construction process which has an important interface with client commissioning (see Chapter 7).

- The relevant consultants identify all services to be commissioned and define the responsibility split for commissioning between designers, contractor, manufacturer and client. Responsibility for specialised plant/services is defined early, particularly 'wear and tear' and the cost of consumables, fuel, power, water, etc.

- The services designers, and commissioning contractor if relevant, audit the final layout drawings to ensure that they make provision for the systems to be commissioned in accordance with the relevant codes of practice.

- The consultants/client, and commissioning contractor if relevant, identify all required statutory and insurance approvals relating to services commissioning, and see that plans are made for meeting requirements and obtaining the approvals (see Appendices C and D of Part 2 of this *Code*).

■ That the client understands the importance of the presence of the client's own maintenance/engineering department/maintenance contractor during the commissioning process

■ That the client considers whether an aftercare engineer needs to be appointed to support the client/user in the first 6–12 months of occupancy.

■ There is a programme showing the timescale and sequence of commissioning and testing and handover events, system by system. This is essential.

■ Arrangements are made to ensure that one person only is responsible for control and management of the client's role in commissioning of services. This could be the client's commissioning officer or the project manager, who should be a member of the client's team as defined in Chapter 7. Although this does not preclude more than one person having the benefit of witnessing the commissioning process.

■ The contract documents *must* make adequate provision for testing, commissioning and performance testing. Confirmation on warranties, defects period, environmental testing at this stage will set out what level of commissioning is required and ensure that responsibility for plant and systems is still with the contractor during an extended commissioning period (see Appendices B and C of Part 2).

Construction and post-construction

■ The consultants must inspect the work for which they have design responsibility, and report on progress and compliance with contract provisions, highlighting any corrective action necessary. A commissioning management specialist may be appointed to carry out much of this work.

■ There must be confirmation that all the contractor's construction programmes include commissioning activities and that they are properly related to preceding construction activities. Activities must be complete, timings reasonable and compatible with planned handover, and properly related to preceding activities.

■ Co-ordination of the consultants' arrangements is required for client involvement in, or observation of, contractor's commissioning against contract arrangements.

■ Monitoring and reporting progress of commissioning will be carried out to ensure that activities start as scheduled and that the requirements for completion before handover are met. Corrective action will have to be initiated as necessary. It is important that commissioning activity durations do not become eroded due to late or incomplete construction work.

■ All 'completed construction' documents should be in place before commissioning an individual system commences, e.g. cleaning out, testing the electrical power and controls to it. Also, the requirements of 'permits to work' and health and safety should be met; and responsibility for insurance should be clearly defined.

■ Statutory/insurance tests should be arranged and undertaken, witnessed by the relevant authority, e.g. building control, utility companies, fire brigade, insurers, etc.

■ Commissioning records, e.g. test results, calibration requirements, certificates and checklists must be properly maintained and copies bound into the operating and maintenance manuals or in separate commissioning manuals to form part of the official handover documentation.

■ Operating and maintenance manuals, 'as-installed' record drawings and the client's staff training have to be provided by the contractor as required under the contract, although it is recommended that these are fully co-ordinated by others, e.g. the commissioning contractor, if appointed.

■ Adopting agreed structure and software for operating and maintenance manuals with copy disks provided for ease of updating.

■ Record drawings being provided in a computer-aided drawing format for ease of updating.

■ Using video recordings during client-training sessions for subsequent repeat visual reference and to assist new maintenance staff advance along the learning curve.

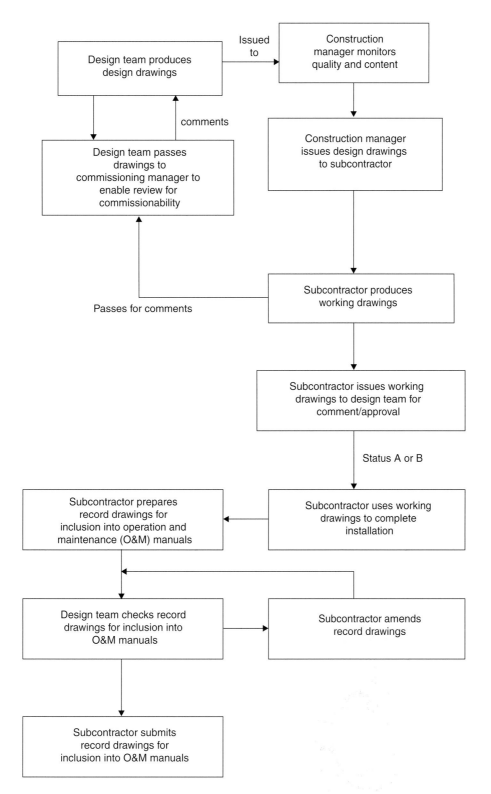

Figure 6.3 Project drawing issue flowchart.

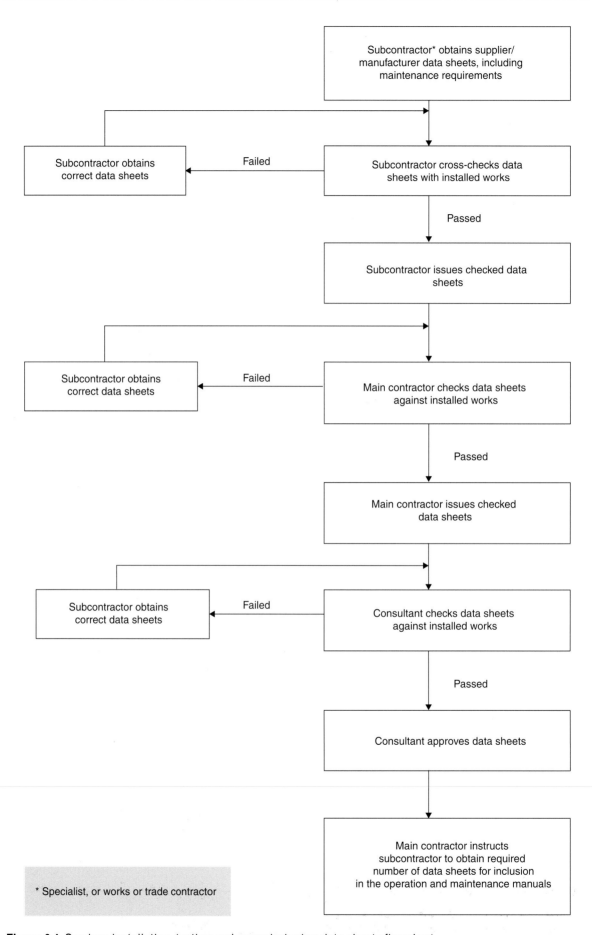

Figure 6.4 Services installation, testing and commissioning data sheets flowchart.

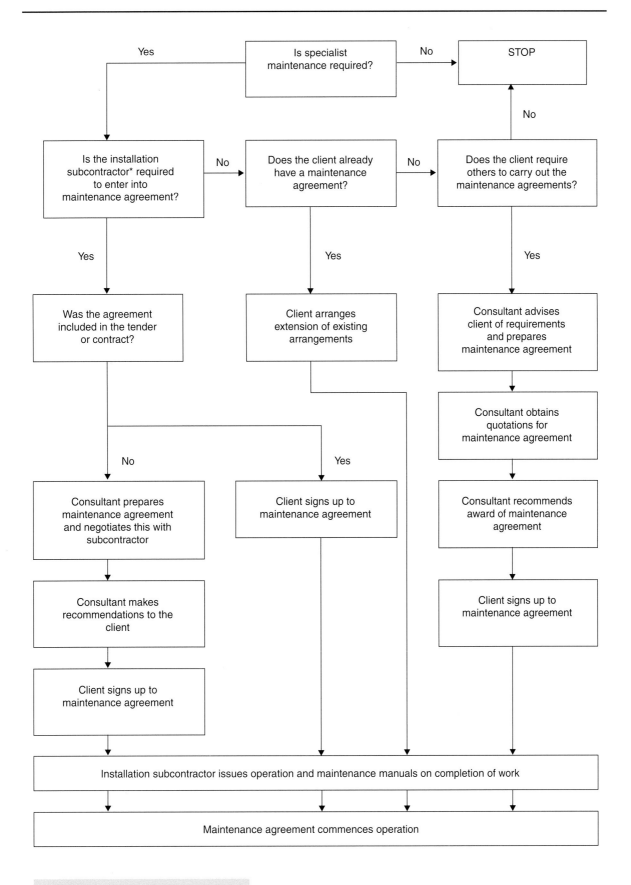

* Specialist, or works or trade contractor

Figure 6.5 Specialist maintenance contracts flowchart.

7

Completion, handover and occupation stage

Client's objectives

At this stage the client's aims include agreeing to a handover plan and schedule and client/supplier responsibilities, especially in terms of criteria for acceptance, provision of necessary project documentation, and defects liability; commissioning arrangements, and any instructions as to future occupation. The client is also to agree and implement handover method and agree a defect rectification plan, if necessary, and a transfer of documentation. Also, an initial post-occupancy review may be undertaken at this stage to highlight any immediate issues for rectification.

Completion

Completion and handover are very much interlinked. This is the final stage of work executed by the contractors and consultants prior to acceptance of the facility by the client. They are carried out under the continued co-ordination and supervision of the project manager, in close working relationship with the consultants. The project manager maintains the required liaison between and acts on behalf of the parties concerned (e.g. the client or the user). Occupation, organised by the client's occupation co-ordinator, is usually preceded by an accommodation schedule of works which can consume anything up to 3% of the construction budget. These works may or may not involve the design consultants and may be managed by the project manager or by the client's accommodation manager.

Project management actions

This stage marks the end of the main construction works, and involves the project manager in a number of activities to successfully terminate the construction contract as follows:

■ Ensuring the contract administrator has inspected the works and, if appropriate, has issued the certificate of practical completion. Attached to the certificate should be a list of outstanding snags and exclusions together with a statement of the timescale for their final completion. The project manager needs to ensure completion of these final items does not cause disruption to the client's use of the end product.

■ The issue of the certificate of practical completion marks a transfer of responsibility for the end product from the contractor to the client. The project manager needs to ensure that the client is prepared for the insurance and security implications of this change of responsibility.

■ A certificate marking the completion of part of the works can be issued at any time during the project. Sectional completion is used for the early handover of part of the end product , e.g. a computer room.

■ Following issue of the certificate of practical completion, the project manager should ensure the final account process for the completion works is concluded with the contractor as quickly as possible. The final account is a reconciliation of the tendered works and the scope of the works finally instructed, and takes account of variations to the contract issued during the course of the project. While assessment of the cost and time implications of these contractual entitlements is initially made by the contract administrator, in the event of the contractor being unhappy at the proposed settlement, the project manager will be asked to arbitrate.

■ The final account process involves consideration of claims for additional monies and time made outside of the contract. The project manager, who will make a recommendation to the client for any awards, will consider these claims for consequential loss. The project manager has a duty to monitor the legal liability of the client throughout the construction work.

■ Ensuring that during the defects liability period there is a system in place for the client to report defects and for the contractor to carry out rectification works. At the end of the defects period, the project manager should ensure that the contract administrator carries out a final inspection and, if appropriate, issues the final certificate.

■ At practical completion a number of significant documents are handed over from the contractor to the client. The project manager needs, on behalf of the client, to ensure firstly, that these documents are available and secondly, that they are to the necessary quality:

  ○ the project's health and safety file

  ○ 'as-built' drawings together with all relevant specifications, etc.

  ○ operating and maintenance manual, consisting of details of maintenance schedules, operating instructions, manufacturer's details

  ○ warranties and guarantees from suppliers.

  ○ copies of statutory authority approvals and consents

  ○ test and commissioning documentation.

Actions by the design team

Design team members should carry out the following actions:

■ Inspect, as appropriate, the work for which they have design responsibility and report to the design team leader, with copy to the project manager, on progress and compliance with contract provisions, highlighting any corrective action to be taken.

■ Inspect work at the practical completion stage, produce the outstanding work schedule and sign off, certifying, subject to completion of works listed in the schedule. As a general rule, a certificate of practical completion should not be issued if there are incomplete or defective works outstanding.

- Modern air-conditioned facilities and control systems require a full range of external temperatures and full occupation to test their adequacy and stability, i.e. summer and winter working.

- Inspect the work at the end of the contract defects liability period, compile defects schedule and subsequently confirm that: (1) all defects have been rectified; (2) any omissions have been made good and (3) all necessary repairs have been carried out.

Planning and scheduling handover

The overall objective is to schedule the required activities to achieve a co-ordinated and satisfactory completion of all work phases within the cost plan. This has to be meshed with the logistical planning of the client's occupation co-ordinator and any accommodation schedule of work to be completed prior to occupation.

Generally, construction projects can be subject to phased (sectional), as well as practical completion. The relevant procedures applied depend on the nature and complexity of the project, and/or requirements of the users. In effect, phased completion means the practical completion for each specific phase of construction. However, this must not:

- prevent or hinder any party from commencing, continuing or completing their contractual obligations

- interfere with the effective operation of any plant or services installations.

In cases of phased completion handover, the user/tenant is usually responsible for insuring the works concerned. On practical completion handover the whole of the insurance premium becomes the user's responsibility.

Procedures

The actual practical completion and handover procedures applicable to a specific project will be detailed by the project manager in the handbook for the project concerned (see Appendix 18 and Appendix E for typical examples). However, the main aspects of completion and handover will generally cover the minutiae of the following activities:

- Preparation of lists identifying deficiencies, e.g. unfinished work, frost damage, and materials, goods, and workmanship not in accordance with standards.

- All remedial and completion work carried out within the specified time under the direct supervision of nominated, qualified and experienced personnel.

- Monitoring and supervising completion and handover against the schedule.

- The provision of the required number of:

  ○ copies of the CDM health and safety file

  ○ 'as-built' and 'installed' record drawings, plans, schedules, specifications, performance data and tests results

  ○ commissioning and test reports, calibration records, operating and maintenance manuals, including related health, safety and emergency procedures

- ○ planned maintenance schedules and specialist manufacturers' working instructions.

■ Monitoring proposals for the training of engineering and other services staff and assistance in the actual implementation of agreed schemes.

■ Ensuring that handover takes place when all statutory inspections and approvals are satisfactorily completed but does not take place if the client/tenant cannot have beneficial use of the facility, i.e. not before specified defects are made good, indicating likely consequences and drawbacks of premature occupation.

■ Setting up procedures to monitor and supervise any post-handover works, which do not form part of the main contract, and to monitor the defects liability period.

■ Initiating, in close co-operation with the relevant consultants, contra-charging measures in cases of difficulties with completing outstanding works or making good any defects.

■ Monitoring progress of final accounts by assisting in any controversial aspects or disputes, and by ascertaining that draft final accounts are available on time and are accurate.

■ Reviewing progress at regular intervals, to facilitate a successful final inspection, and the issuing of a final certificate.

■ Establishing the plan for post-completion project evaluation and feedback from the parties to the contract for the post-completion review project close-out report.

Client commissioning and occupation

Having accepted the constructed structure from the contractor at practical completion, the client has to finally prepare the facilities ready for occupation. This stage of the project lifecycle comprises three major groups of tasks: client accommodation works, operational commissioning and migration.

■ In order to allow as much time as possible for the client organisation to develop their detailed requirements, or to reflect their latest business 'shape', it is common for the client to organise a further project to carry out accommodation works. It is likely the project manager will be involved to manage the project team established to carry out these works. Often this team will be separate from the main project team and will comprise personnel with greater experience of operating in a finished project environment.

Typical elements of client accommodation works for an office building would be:

- ○ Fitting out of special areas
  - – restaurant/dining areas
  - – reception areas
  - – training areas
  - – executive areas
  - – post rooms
  - – vending areas.

- ○ Installation of IT systems

  - – servers

  - – desktop PCs

  - – telecommunications equipment

  - – fax machine

  - – audiovisual and video conferencing.

- ○ Demountable office partitions

  - – furniture

  - – specialist equipment

  - – security systems

  - – artwork and planting.

Operational commissioning

The principles of client commissioning and occupation should be determined at the feasibility and strategy stage. Client commissioning (as with occupation, which usually follows on as a continuous process) is an activity predominantly carried out by the client's personnel, assisted by the consultants as required.

The objective of client commissioning is to ensure that the facility is equipped and operating as planned and to the initial concept of the business plan established for the brief. This entails the formation, under the supervision of the client's occupation co-ordinator, of an operating team early in the project so that requirements can be built into the contract specifications. Ideally, the operating team is formed in time to participate in the design process. (Their role is identified in Part 2, supported by a checklist in Appendix F.)

Main tasks

The main tasks are as follows:

- ■ Establishing the operating and occupation objectives in time, cost, quality and performance terms. Consideration must be given to the overall implications of phased commissioning and priorities defined for sectional completions, particular areas/services and security.

- ■ Arranging the appointment of the operating team in liaison with the client. This is done before or during the detailed design stage, so that appropriate commissioning activities can be readily included in the contract.

- ■ Making sure at budget stage that an appropriate allowance for the client's commissioning costs is made. Accommodation schedule of works can potentially consume a significant part of the total project budget.

- ■ Preparing role and job descriptions (responsibilities, time-scales, outputs) for each member of the operating team. These should be compatible with the construction programme and any other work demands on members of the operating team.

- ■ Co-ordinating the preparation of the client's commissioning schedule and an action list in liaison with the client, using a commissioning checklist (see Appendix F).

■ Arranging appropriate access, as necessary, for the operating team and other client personnel during construction, by suitable modification of the contract documents.

■ Arranging co-ordination and liaison with the contractors and the consultants to plan and supervise the engineering services commissioning, e.g. preparation of new work practices manuals, staff training and recruitment of additional staff if necessary; the format of all commissioning records; renting equipment to meet short-term demands; overtime requirements to meet the procurement plan; meeting the quality and performance standards, all as defined in Chapter 6.

■ Considering early appointment/secondment of a member of the client management team to act as the occupation co-ordinator; this ensures a smooth transition from a construction site to an effectively operated and properly maintained facility (see Appendix 19 for an introduction to facilities management).

■ Before the new development can be occupied, the client needs to operationally commission various elements of the development. This involves setting to work various systems and preparing staff ready to run the development and its installations:

   ○ transfer of technology

   ○ checking voice and data installation are operational

   ○ stocking and equipping areas such as a restaurant

   ○ training staff for running various systems

   ○ training staff to run the property.

■ Also part of the client's operational commissioning is the obtaining of the necessary statutory approvals needed to occupy the building, such as the occupation certificate and the environmental health officer's approval of kitchen areas (if applicable).

■ Occupation of developed property depends on detailed planning of the many spaces to be used. For office buildings this space planning process is developed progressively throughout the project lifecycle.

Final determination of seating layouts is delayed until the occupation stage in order to accommodate the latent changes to the client's business structure. A typical space planning process consists of:

   ○ confirming the client's space standards including policy on open plan and cellular offices

   ○ confirming the client's furniture standards

   ○ determining departmental headcount and specific requirements

   ○ determining an organisational model of the client's business, reflecting the operational dependencies and affinities

   ○ develop a building stacking in order to fit the gross space of each department within the overall space of the building

   ○ develop departmental layouts to show how each department fits the space allocated to it

○   develop furniture seating layouts in order to allocate individual names to desks

It is essential that for each of these stages the client organisation in the form of user liaison groups has a direct involvement and approves each stage.

■   Moving or combining businesses into new premises is a major operation for a client. During the duration of the move there is the potential for significant disruption to the client's business. The longer the move period, the greater the risks to the client. Migration therefore requires a significant level of planning. Often the client will appoint a manager separate from the new building project to take overall responsibility for the migration. For major or critical migrations, the client should consider the use of specialist migration consultants to support their own resources.

■   During the migration planning a number of key strategic issues need to be addressed:

○   determining how the building will be occupied

○   establishing the timing of the move

○   identifying the key activities involved in the migration and assigning responsible managers

○   determining move groups and sequence of moves to minimise business disruption

○   determining the project structure for managing the move

○   identifying potential risks that could impact on the move

○   involving and keeping the client's staff informed.

As some of these strategic issues could have an impact on the timing and sequencing of the main building works, it is important to address them early in the project lifecycle.

■   The final part of occupation is the actual move management. This involves the appointment of a removal contractor, planning the detailed tactics of the move, and supervision of the move itself.

■   The overall period that the move takes is determined by the number of items to be transferred with each member of the staff and by the degree of difficulty of transferring IT systems for each move group.

■   A critical decision for the client during the occupation stage is the point at which a freeze is imposed on space planning and no further modifications are accommodated until after migration has been achieved.

It is likely that the factor having most impact on the timing of the freeze date will be the setting up of individual voice and data system profiles.

It would be common for clients to impose an embargo on changes both sides of the migration and for the client then to carry out a post-migration subproject to introduce all the required changes required by departments.

Client occupation

Occupation should follow a very carefully planned logistical schedule managed by the incoming user of the facility following completion of construction. This can be put under the overarching control of the project manager or can be headed by an appointed occupation co-ordinator.

Unlike many other project management activities, occupation involves employees themselves and is affected by the style of management and the culture of the user's organisation. Consequently, well-executed planning and the end user's involvement in the process can result in better management/employee relations, bringing a greater feeling of participation and commitment to the workforce.

It is normal to produce an operational policy document in the planning stages which is a blueprint for implementation of the business plan. In particular it sets out services which will be kept in house and those services which will be contracted and how they will be procured.

The arrangements for occupation and migration from one facility to another will, on many projects, be predetermined by space-planning exercises carried out in the initial design stages by the design team or space-planning consultant. The guidance given in this chapter should be put in the context of the overall planning of a client's needs for a particular facility. This will follow from:

■ strategic analytical briefing

■ detailed briefing (departmental level)

and lead to criteria such as:

■ quantifying spatial requirements

■ physical characteristics for each department/sector

■ critical affinity groupings

■ extent of amenities

■ workspace standards

■ office automation strategies

■ security/public access

■ furniture, fittings and equipment (FF&E) schedules.

On complex projects this can be taken one stage further to the production of room data sheets which form the basis of the design brief, equipment transfer or purchase, movement of personnel and facilities management.

The procedure outlined below gives a typical approach, which may need to be interpreted in order to harmonise with the practices and expectations of the users. Nevertheless, change in established practices is encouraged where doing so will smooth the process and make it more effective. Occupation can be divided into four stages as explained below, and as shown in Figures 7.1–7.4. The following services are often outsourced on renewable annual or 3-year contracts:

■ reception and telephony

■ security

■ cleaning

■ building management and operation of services and equipment

■ maintenance

■ IT support

Part 1 Project management

- catering and waste management
- landscaping and grounds maintenance
- transport and courier services.

Structure for implementation

Structure for implementation means appointment of individuals and groups to set out the necessary directions, consultation and budget/cost parameters. Figure 7.1 gives an example.

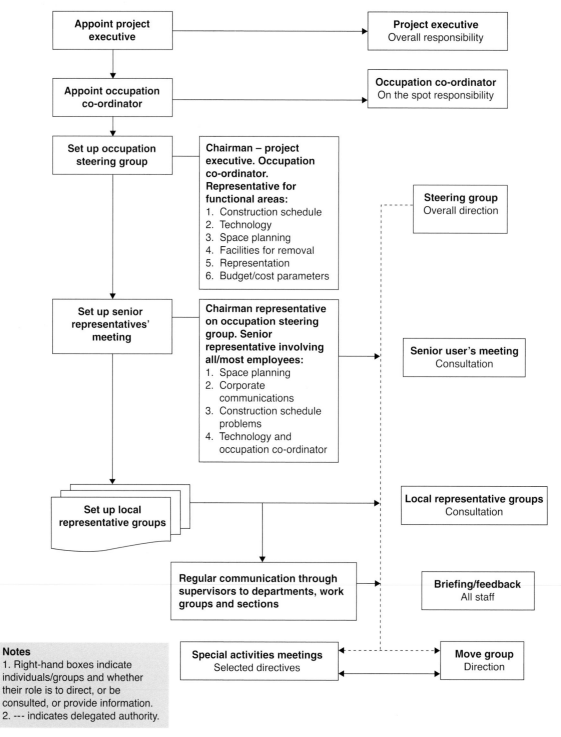

Figure 7.1 Occupation: structure for implementation.

Scope and objectives

Scope and objectives means deciding what is to be done, considering the possible constraints and reviewing as necessary. Figure 7.2 gives an example.

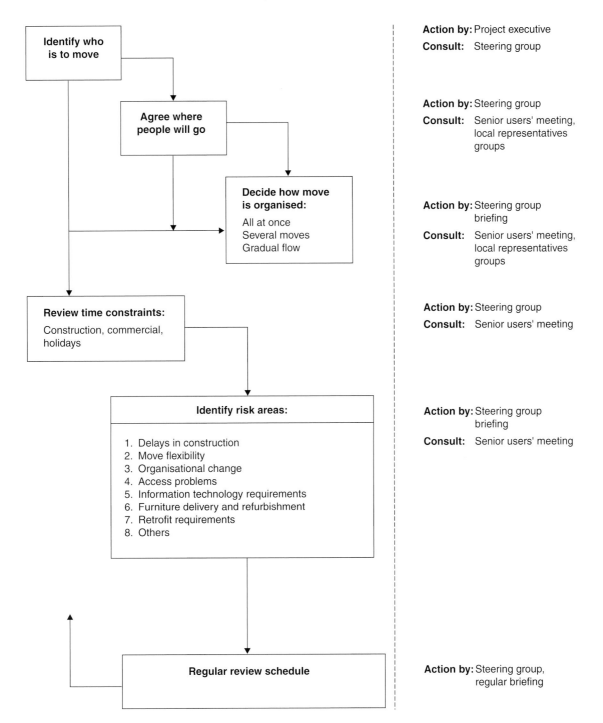

Figure 7.2 Occupation: scope and objectives.

Methodology

Methodology is how the whole process will be achieved. Identification of individual or groups of special activities and their task lists aimed at defining the parameters and other related matters, e.g. financial implications. Figure 7.3 gives an example.

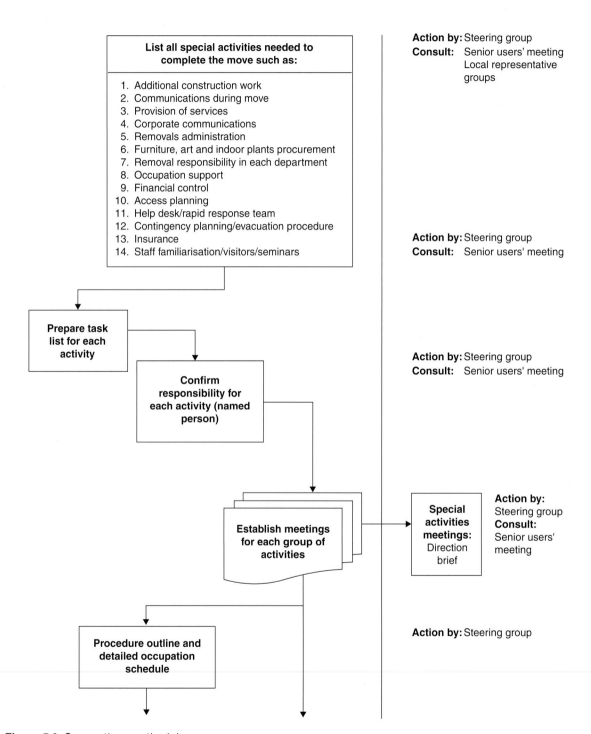

Figure 7.3 Occupation: methodology.

Organisation and control

Organisation and control means carrying out the process and keeping schedule and budget/cost under review. Figure 7.4 gives an example.

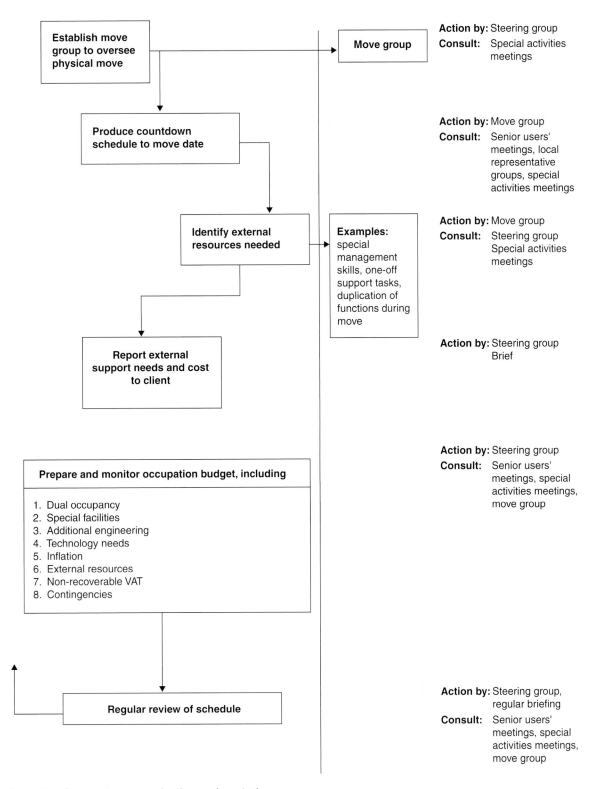

Figure 7.4 Occupation: organisation and control.

The individuals and groups likely to be concerned are as follows:

- *Project executive*: appointed by the client/tenant at the director/senior management level and responsible for the complete process.

- *Occupation co-ordinator*: project manager appointed, or existing one confirmed by the client, with on-the-spot responsibility.

- *Occupation steering group*: chaired by the project executive and consisting of occupation co-ordinator and a few selected senior representatives covering the main functional areas. Concerned with all major decisions but subject to any constraints laid down by the client, e.g. financial limits.

- *Senior representatives meeting*: chaired by one of the functional representatives on the occupation steering group and made up of a few senior representatives covering the majority of employees and the occupation co-ordinator.

- *Local representative groups*: chaired by manager/supervisor of own group and concerned with providing views related to a particular location or department. Membership to reflect the specific interest of the group at the location.

- *Special activities meetings*: meetings for individual or group of special activities as identified in 'methodology'. A single person will be made responsible for achieving all the tasks which make up a special activity and will chair the respective meetings.

- *Move group*: responsible for the overall direction of the physical move, having been delegated by the occupation steering group, the task of detailed preparation and control of the move programme including its budget/cost.

- *Briefing groups*: concerned with effective and regular communication with all employees to provide information to work groups/sections by their own managers or supervisors, so that questions for clarification are encouraged. Special briefings may also be vital, especially during the build-up to occupation.

On many projects, the individuals and groups identified above may be synonymous with those given under client commissioning, e.g. for commissioning team read occupation steering group and vice versa.

Figures 7.1–7.4 provide an at-a-glance summary of the occupation process and Appendices C–F of Part 2 provide of provides checklists for a typical control system.

8 Post-completion review/project close-out report stage

Client's objectives

The client's aims at the closing stage of a project should include:

■ To measure performance of all aspects of the project and ensure that the value of the knowledge gained can be carried forward to future projects.

■ To undertake an initial assessment of the new facility so as to establish its fitness for purpose and satisfaction of requirements.

Introduction

The objective of the review is to make a thorough assessment of all elements of the project and to draw out or feed back, for the benefit of the client, the project management practice concerned and other team members, any lessons and conclusions for application to future projects, i.e. what could have been done differently to mutual advantage? This review/report is good practice but should not be regarded as mandatory and may not be required by all clients. It is worth involving the design consultants if only to check that the client is getting maximum use out of the facilities provided and, in particular, that operational costs are at an optimum level. The typical review is likely to consist of the following elements.

Project audit

■ Brief description of the objective of the project.

■ Summary of any amendments to the original project requirements and their reasons.

■ Brief comment on project form of contract and other contractual/agreements provisions. Were they appropriate?

■ Organisation structure, its effectiveness and adequacy of expertise/skills available.

■ Master schedule: project milestones and key activities highlighting planned versus actual achievements.

■ Unusual developments and difficulties encountered and their solutions.

■ Brief summary of any strengths, weaknesses and lessons learned, with an overview of how effectively the project was executed with respect to the designated requirements of:

- cost
- scheduling and programming
- technical competency
- quality
- health and safety aspects
- sustainability targets (environmental, social and economic).

■ Was the project brief fulfilled and does the facility meet the client/user needs? What needs tweaking and how could further improvements be made on a value-for-money basis?

■ Indication of any improvements that could be made in future projects.

Cost and time study

■ Effectiveness of:

- cost and budgetary controls
- claims procedures.

■ Authorised and final cost.

■ Planned against actual costs (e.g. S-curves) and analysis of original and final budget.

■ Impact of claims.

■ Maintenance of necessary records to enable the financial close of the project.

■ Identification of time extensions and cost differentials resulting from amendments to original requirements and/or other factors.

■ Brief analysis of original and final schedules, including stipulated and actual completion date; reasons for any variations.

Human resources aspects

■ Communication channels and reporting relationships (bottlenecks and their causes).

■ Industrial relations problems, if any.

■ General assessment and comments on staff welfare, morale and motivation.

Performance study

■ Planning and scheduling activities.

■ Were procedures correct and controls effective?

■ Staff hours summary:

- breakdown of planned against actual
- sufficiency of resources to carry out work in an effective manner.

■ Identification of activities performed in a satisfactory manner and those deemed to have been unsatisfactory.

- Performance rating (confidential) of the consultants and contractors, for future use.

Project feedback

Project feedback necessarily reflects the lessons learnt at various stages of the project, including recommendations to the client for future projects. Ideally feedback should be obtained from all of the participants in the project team at various stages. If necessary, feedback can be obtained at the key decision-making stage (e.g. at the completion of each of the seven stages as outlined in this *Code of Practice*). The project feedback form should include:

- brief description of the project

- outline of the project team

- form of contract and value

- feedback on contract (suitability, administration, incentives, etc.)

- technical design

- construction methodology

- comments on the technical solution chosen

- any technical lessons to be learnt

- form of consultant appointments

- comments on consultant appointments

- project schedule

- comments on project schedule

- cost plan

- comments on cost control

- change management system

- values of changes

- major source(s) of changes/variations

- overall risk management performance

- overall financial performance

- communication issues

- organisational issues

- comments on client's role/decision-making process

- comments on overall project management including any specific issues

- other comments

- close-out report.

It has to be remembered that the purpose of a project feedback is not just only to express what went wrong and why, but also to observe what has been achieved well, and if (and how) that can be improved in future projects, i.e. continuous improvement.

Close-out report

The project manager should summarise the findings from the various post-completion reviews in a close-out report which is issued to the client as a formal record of the project's delivery and outcome.

Benefits Realisation

Although a client responsibility, on some projects the project manager may be required to assist in the assessment of whether the intended benefits for the client's organisation have been realised.

PART 1 APPENDICES

APPENDIX 1 Typical terms of engagement

Job title: Project manager.

Date effective:

General objective

Acting as the client's representative within the contractual terms applicable, to lead, direct, co-ordinate and supervise the project in association with the project team.

The project manager will ensure that the client's brief, all designs, specifications and relevant information are made available to, and are executed as specified with due regard to cost by, the design team, consultants and contractors (i.e. the project team), so that the client's objectives are fully met.

Relationships

Responsible and reporting to   The client.

Subordinates   Practice support staff and secretarial/clerical staff.

Functional   Fully integrated working with any project support staff who are not line subordinates:

○   liaison, as required/expedient with relevant client's staff, e.g. legal, insurance, taxation

○   full interdependent co-operation with:

(a)   design team and consultants

(b)   contractors.

External   Liaison with local or other relevant authority on matters concerning the project.

Contact with suppliers of construction materials/equipment, in order to be aware of the most efficient and cost-effective application, and working methods.

Contact with:

■   Client's information and communication technology (ICT) team or other higher technology sources, able to provide expertise on the application of advanced technology in the design and/or construction processes of the project (e.g. communications, environment, security and fire prevention/ protection systems).

■   And preferably, membership of appropriate professional bodies/societies.

Authority   The definition of the authority of the project manager is a key requirement in enabling him to manage the successful achievement of the client's objectives. The extent must be clearly defined. A distinction should be drawn between the responsibility that the project manager may have which concerns his accountability for different aspects of the project, and the authority which will determine the ability of the project manager to control, command and determine the commitment of resources to the project. The full extent of the responsibility and authority vested in the project manager will depend on the terms and duties included in the project management agreement.

The extent of the project manager's responsibility and authority may be balanced, but the two may be unequal. Frequently, the project manager may have extensive responsibility in an area that does not carry commensurate authority, or vice versa.

The authority of the project manager should be defined regarding his obligations to issue instructions, approve limits of expenditure, and when to notify the client and seek the instructions of the client in matters relating to:

■ the schedule and time taken to complete the project

■ expenditure and costs, including development budget, project cost plan, and financial rewards and viability

■ designs, specifications and quality

■ function

■ contractors' contracts

■ consultants' appointments

■ assignment of contracts or appointments

■ administrative procedures, including issuing or signing of correspondence, certification and other project documentation.

The client and the project manager should give careful consideration to the authority that will be necessary to ensure the successful achievement of the client's objectives and, if necessary, establish appropriate lines of authority and communication within the client organisation to facilitate the implementation of agreed procedures.

Detailed responsibilities and duties

1. Analysis of the client's objectives and requirements, assessment of their feasibility and assistance in the completion of project brief and establishment of the capital budget.

2. Formulation, for the client's approval, of the strategic plan for achieving the stated objectives within the budget, including, where applicable, the quality assurance scheme.

3. Generally keeping the client informed, throughout the project, on progress and problems, design/budgeting/construction variations, and such other matters considered to be relevant.

4. Participation in making recommendations to the client, if required, in the following areas:

■ The selection of the consultants as well as in the negotiation of their terms and conditions of engagement.

■ The appointment of contractors/subcontractors, including the giving of advice on the most suitable forms of tender and contract.

5. Preparation for the client's approval of the following items:

■ The overall project schedule embracing site acquisition, relevant investigations, planning, pre-design, design, construction and handover/occupation stages.

- Proposals for architectural and engineering services. The project manager will monitor progress and initiate appropriate action on all submissions concerned with planning approvals and statutory requirements (timely submission, alternative proposals and necessary waivers).

- The project budget and relevant cash flows, giving due consideration to matters likely to affect the viability of the project development.

6. Finalisation of the client's brief and its confirmation to the consultants. Providing the client with all existing and, if necessary, any supplementary data on surveys, site investigations, adjoining owners, adverse rights or restrictions and site accessibility/traffic constraints.

7. Recommending to the client and securing approval for any modifications or variations to the agreed brief, approved designs, schedules and/or budgets resulting from discussions and reviews involving the design team and other consultants.

8. Setting up the management and administrative structure for the project and thereby defining:

- responsibilities and duties, as well as lines of reporting, for all parties

- procedures for clear and efficient communication

- systems and procedures for issuing instructions, drawings, certificates, schedules and valuations and the preparation and submission of reports and relevant documentary returns.

9. Agreeing tendering strategy with the consultants concerned.

10. Advising the client as necessary on the following items:

- The progress of the design and the production of required drawings/ information and tender documents, stressing at all times the need for a cost-effective approach to optimise costs in construction methods, subsequent maintenance requirements, preparation of tender documents and performance/workmanship warranties.

- The correctness of tender documents.

- The prospective tenderers pre-qualified by the design team and other consultants involved, obtaining additional information if pertinent and confirming accepted tenders to the client and the consultants.

- The preliminary construction schedule for the main contractors, agreeing any revisions to meet fully the client's requirements and releasing this to the project team for action.

- The progress of all elements of the project, especially adherence to the agreed capital and sectional budgets, as well as meeting the set standards and initiating any remedial action.

- The contractual activities the client must undertake, including user study groups and approval/decision points.

11. Establishing with the quantity surveyor the cost monitoring and reporting system and providing feedback to the other consultants and the client on budget status and cash flow.

12. Organising and/or participating in the following activities:

- Presentations to the client, with advice on and securing approval for the design of fabrics, finishes, fitting-out work and the environment of major interior spaces.

- All meetings with the project team and others involved in the project (chairing or acting as secretary) to ensure:

  ○ an adequate supply of information/data to all concerned

  ○ that progress is in accordance with the schedule

  ○ that costs are within the budgets

  ○ that required standards and specifications are achieved

  ○ that contractors have adequate resources for the management, supervision and quality control of the project

  ○ that the relevant members of the project team inspect and supervise construction stages as specified by the contracts.

13. Responsible for:

- preparation of the project handbook

- achieving good communications and motivating the project team

- monitoring progress, costs and quality and initiating action to rectify any deviations

- setting priorities and effective management of time

- co-ordinating the project team's activities and output

- monitoring project resources against planned levels and initiating necessary remedial action

- preparing and presenting specified reports to the client

- submitting time sheets and other data on costing and control to the client

- processes, including required returns and all other relevant information

- approving, in collaboration with the project team and within the building contract provisions, any sublet work

- identifying any existing or potential problems, disputes or conflicts and resolving them, with the co-operation of all concerned in the best interests of the client

- recommending to the client the consultants' interim payment applications and monitoring such applications from contractors

- monitoring all pre-commissioning checks and progress of any remedial defects liability work and the release of retention monies

- verifying with the project team members concerned any claims for extensions of time or additional payments and advising the client accordingly

- checking consultants' final accounts before payment to the client

- monitoring the preparation of contractors' final accounts, obtaining relevant certificates and submitting them for settlement by the client

■ ensuring the inclusion in the contract and subsequently requesting the design team, consultants and contractors to supply the client with as-built and installed drawings, operating and maintenance manuals, and health and safety file, as well as ensuring arrangements are made for effective training of the client's engineering and maintenance staff, i.e. facilities management.

14. Taking all appropriate steps to ensure that site contractors and other regular or casual workers observe all the rules, regulations and practices of safety and fire prevention/protection. Exercising 'good site housekeeping' at all times.

15. Participating in the final cost reconciliation or final account of the project and taking such action as directed or required.

Extra-project activities

Participating in informal discussions with own and other practices, as well as the client's staff, on technical details, methods of operations, problem solving and any other pertinent actions relevant to present or previous projects, in order to exchange views/knowledge conducive to providing a more effective overall performance.

The project manager has responsibility for the following areas:

■ Personnel matters relating to his staff, including appraisal/reviews, training/development and job coaching and counselling, as defined by the client and/or project management practice guidelines and procedures.

■ Updating himself and staff in new ideas relating to project management, including management/supervisory skills and practice generally, business, financial, legal and economic trends, the latest forms of contract, planning and Building Regulations, as well as advances in construction techniques, plant and equipment.

Terms of engagement: the services contracts

1. The CIC Consultants Contract Conditions and Scope of Services 2007.

2. RICS Project Management Agreement (3rd edn, 1999)

3. APM Terms of Appointment for a Project Manager (1998)

4. NEC3 Professional Services Contract (PSC) (2005)

5. RIBA Form of Appointment for Project Managers (2004)

6. NHS Estates Agreement for the Appointment of Project Managers for commissions for construction projects in the NHS

APPENDIX 2   Health and safety in construction including CDM guidance

Generally, the laws governing health and safety relate to all construction activities (including design) and are not industry specific. There are several Acts and Regulations involved and information on how to access these is provided at the end of this appendix.

Some of the principal Acts which deal with health, safety and welfare in construction are as follows:

- Health and safety at Work etc. Act 1974

- Mines and Quarries Act 1954

- Factories Act 1961

- Offices, Shops and Railways Premises Act 1963

- Employers Liability Acts – various

- Control of Pollution Act 1989

- Highway Act 1980

- New Roads and Streetworks Act 1991

- Corporate Manslaughter and Corporate Homiside Act 2007

The fundamental Act governing health and safety in construction is the Health and safety at Work etc. Act 1974. This act has some 62 separate Regulations and it is not possible to deal with such a large subject area here, however, the principal regulations of this Act, which affect design and construction, are:

- Management of Health and Safety at Work Regulations 1999 amended 2006

- Construction (Design and Management) Regulations 2007 (known as the CDM Regulations)

- The Work at Height Regulations 2005 amended 2007.

Some other related regulations and guides are:

- Site Waste Management Plans Regulations 2008

- Reporting of Injuries, Diseases and Dangerous Occurrences Regulations (RIDDOR) 1995

- The Control of Major Accident Hazards Regulations 1999 (COMAH) amended 2005

- The Chemicals (Hazard Information and Packaging for Supply) Regulations 2003 (CHIP 3)

- The Health and Safety (Display Screen Equipment) Regulations 1992

- COSHH (Control of Substances Hazardous to Health) Regulations 2002: Provision and Use of Work Equipment Regulations (PUWER 98)

- Lifting Operations and Lifting Equipment Regulations (LOLER 98)

- Personal Protective Equipment at Work Regulations 1992

- Signposts to the Health and Safety (Safety, Signs and Signals) Regulations 1996

- Control of Asbestos Regulations 2006

CDM 2007 Regulations

The Construction (Design and Management) Regulations 1994 is a Statutory Instrument that was introduced in compliance with the European Directives signifying a major change in the focus of the industry in terms of management of health and safety. For the first time, specific and explicit duties were placed on the client and the designers in context of management of health and safety. These regulations were updated in 2007.

The 2007 CDM Regulations have been introduced to provide a simplified regulatory structure with greater focus on planning and management of construction activities, with strengthened requirements on co-operation and co-ordination among all the key parties.

CDM 2007 contains five parts:

- Part 1 – introduction

- Part 2 – general management duties applying to all construction projects

- Part 3 – additional duties where projects are notifiable

- Part 4 – duties relating to health and safety on construction sites

- Part 5 – general.

In addition, it is also supported by a CDM 2007 Approved Code of Practice (ACoP).

These regulations apply to all construction work, regardless size and duration. However, the 'notifiable' projects trigger additional duty holders and duties as identified in Part 3 of the regulations including:

- principal contractor

- CDM co-ordinator (the role of planning supervisor under CDM 1994 has been removed)

- notification to the Health and Safety Executive (HSE) (F10 form)

- construction phase plan

- health and safety file.

Regardless of notification, most duties remain on clients, designers and contractors, and in most cases, there are additional duties.

Notifiable projects

The triggers for notification have also been simplified from CDM 1994. Under CDM 2007, notifiable construction projects are with a non-domestic client and involve

Part 1 Appendices

either construction work lasting longer than 30 days or construction work involving 500 person-days.

Definition of client

CDM 2007 defines a client as an individual or an organisation who, in the course or furtherance of a business, has a construction project carried out by another or by themselves. This excludes domestic clients (i.e. someone who lives or will live in the premises where the work is carried out). The CDM client duties will still apply to domestic premises if the client is a local authority, landlord, housing association, charity, collective of leaseholders or any other trade or business.

For PFI or PPP projects, the project originators are the client at the start of the project until the special-purpose vehicle has been set up and has assumed the role of the client.

Role of clients under CDM 2007

The 2007 CDM Regulations do not, in the main, place new duties on the client, however:

- Existing duties under the old CDM regulations, as well as other relevant regulations, have been made explicit.

- Clarifications have been made with respect to how these duties should be exercised.

- Clients are accountable for the impact they have on health and safety.

- Clients are to ensure that a CDM co-ordinator is appointed to advise and co-ordinate activities on notifiable projects and adequate time and resources are provided to allow the project to be delivered safely.

- It is the responsibility of the client to provide key information to the designers and contractors; also it is for the client to arrange for any gaps in information to be filled in (e.g. commissioning an asbestos survey).

- Clients are to appoint a competent CDM co-ordinator and a competent principal contractor, and to ensure that the construction phase does not start unless the welfare facilities are in place and the construction phase health and safety plan is prepared.

- It is the responsibility of the client to retain and provide access to the health and safety file and revise it with any new information.

- For notifiable projects where no CDM co-ordinator or principal contractor is appointed, the client will be deemed to be the CDM co-ordinator and/or the principal contractor and be subject to their duties.

If the client makes a reasonable judgement that the contractor's management arrangements are suitable, taking account of the nature and risks of the project, and it is clearly based on evidence, clients will not be criticised if the arrangements subsequently prove to be inadequate or fail to be implemented without the client's knowledge.

Role of the CDM co-ordinator

CDM 2007 has created the new role of CDM co-ordinator which replaces the role of the planning supervisor under CDM 1994. The CDM co-ordinator is to advise the client on health and safety issues during design and planning phases of construction work and is only need to be appointed for notifiable projects. The key duties include:

- Advising the client about selecting competent designers and contractors – they do not have to approve the appointments.

- Helping to identify what information will be needed by designers and contractors.

- Co-ordinate the arrangements for health and safety of planning and design work – they do not have to check the designs, although they have to be satisfied that the hierarchy is addressed.

- Ensuring that the HSE is notified of the project (for notifiable projects only).

- Advising on the suitability of the initial construction phase plan – they do not have to approve or supervise the principal contractor's construction phase plan or work on site or approve risk assessments and method statements.

- Preparing a health and safety file.

The duties of a CDM co-ordinator can be carried out by a client, principal contractor, contractor, designer or a full-time CDM co-ordinator.

Role of designers

Under CDM 2007 a designer is someone who designs or specifies building work, including non-notifiable and domestic projects. This includes people who prepare drawings, design details, analysis and calculations, specifications and bills of quantities. In broader terms, these would include civil and structural engineers, building services engineers, material specifiers, temporary works designers, interior fit-out designers, clients who specify and design and build contractors. However, local authorities, etc. providing advice on relevant statutory requirements are not designers. If they require particular features that are not statutory then they are designers.

The key duties of the designers, under CDM 2007, are:

- Ensure clients are aware of their duties.

- Make sure they (the designer) are competent for the work they do.

- Co-ordinate their work with others as necessary to manage risk.

- Co-operate with CDM co-ordinator and others.

- Provide information for the health and safety file.

- Eliminate hazards from the construction, cleaning, maintenance, and proposed use (workplace only) and demolition of a structure.

- Reduce risks from any remaining hazard.

- Give collective risk reduction measures priority over individual measures.

- Take account of the Workplace (Health, Safety & Welfare) Regulations 1992 when designing a workplace structure.

- Provide information with the design to assist clients, other designers and contractors.

- In particular inform others of significant or unusual or 'not obvious' residual risks.

- Risks which are not foreseeable do not need to be considered.

- CDM 2007 does not require 'zero-risk' designs.

- The amount of effort made to eliminate hazards should be proportionate to the risk.

- Check that the client has appointed a CDM co-ordinator.

- Only 'initial' design work is permitted until a CDM co-ordinator has been appointed: initial design can be considered to be no more than:

    ○ work within and beyond RIBA Stage C

    ○ work within and beyond CIC Consultant Contract 2006 Stage 3 (draft as at August 2006)

    ○ work beyond OGC Gateway 1

    ○ work within and beyond ACE Agreement A(1) or B (1) 2002 Stage C3.

- Co-operate with the CDM co-ordinator, principal contractors and with other designers or contractors so all can comply with their CDM duties.

- Provide relevant information for the health and safety file.

Role of the principal contractor

The roles and responsibilities for the principal contractor have in fact been changed very little between CDM 1994 and CDM 2007. A principal contractor must be appointed as soon as practicable for all notifiable project and the principal contractor should ensure that client is aware of appropriate duties, that the CDM co-ordinator has been appointed and that the HSE notified. Further responsibilities of the principal contractor include:

- Those they appoint, including all contractors and subcontractors, are competent.

- That the construction phase is properly planned, managed, monitored and resourced.

- That contractors are made aware of the minimum time allowed for planning and preparation.

- Providing relevant information to contractors.

- Ensuring safe working, co-ordination and co-operation between contractors.

- Ensuring that the construction phase health and safety plan is prepared and implemented. This plan needs to set out the organisation and arrangements for managing risk and co-ordinating work and should be tailored to the particular project and risks involved.

- Making sure that suitable welfare is available from the start of the construction phase.

- Ensuring that site rules as required are prepared and enforced.

- Providing eeasonable direction to contractors including client-appointed contractors.

- Controlling access to the site to restrict unauthorised entry.

- Making the construction phase plan available to those who need it.

- Providing information promptly to the CDM co-ordinator for the health and safety file.

- Liaising with the CDM co-ordinator in relation to design and design changes.

- Ensuring all workers have been provided with suitable health and safety induction, information and training.

- Ensuring that the workforce is consulted about health and safety matters.

Under CDM 2007 a principal contractor does not have to:

- Provide training to workers they do not employ (it is the responsibility of individual contractors to train those they employ)

- undertake detailed supervision of contractors' work.

Duties on contractors and self-employed workers

- Check clients are aware of their duties.

- Plan, manage and monitor their own work to make sure that their workers are safe.

- Ensure that they and those they appoint are competent and adequately resourced.

- Inform any contractor that they engage, of the minimum amount of time they have for planning and preparation.

- Provide their workers (whether employed or self-employed) with any necessary information and training and induction.

- Report anything that they are aware of that is likely to endanger the health and safety of themselves or others.

- Ensure that any design work they do complies with CDM designer duties.

- Comply with the duties for site health and safety.

- Co-operate and co-ordinate with others working on the project.

- Consult the workforce.

- Not begin work unless they have taken reasonable steps to prevent unauthorised access to the site.

- Obtain specialist advice (e.g. from a structural engineer or occupational hygienist) where necessary.

- Check that a CDM co-ordinator has been appointed and that have been HSE notified before they start work (for notifiable projects only).

- Co-operate with the principal contractor, CDM co-ordinator and others working on the project.

- Inform the principal contractor about risks to others created by their work.

- Comply with any reasonable directions from the principal contractor.

- Work in accordance with the construction phase plan.

- Inform the principal contractor of the identity of any contractor they appoint or engage.

- Inform the principal contractor of any problems with the plan or risks identified during their work that have significant implications for the management of the project.

- Inform the principal contractor about any death, injury, condition or dangerous occurrence.

- Provide information for the health and safety file.

Duties to control site health and safety

Part 4 of CDM 2007 contains the duties to control specific construction health and safety risks. It has similar duties compared to the old Construction (Health, Safety and Welfare) Regulations 1996, which Part 4 replaces, in that:

- Applies to all construction sites.

- Duties are on every contractor and every other person who controls construction work.

- The wording and style has been updated and structure altered in parts, but retains most of the basic requirements of the original regulations.

There are some changes however:

- Good order now requires a site to be identified by suitable signs, be fenced off or both in accordance with the level of risk.

- New requirement to record in writing arrangements for demolition and dismantling.

- Excavation, cofferdam and caisson provisions have been extensively rewritten to make them more succinct and cohesive.

- Duties on reports and inspections have been restructured.

- Rest facilities, now requires seats with backs (specific requirement of the European directive – only required if replacing existing seating).

- Training and competence, specific requirements covered in the general part of the regulations.

- Requirements for doors and gates have been moved to Workplace (Health, Safety & Welfare) Regulations 1992.

- The provision to have equipment available to replace a rail vehicle on to its tracks has been removed.

- Prevention of drowning, provisions on a vessel's construction, maintenance and under control by competent person have been removed.

Competence and training

To be competent, an organisation or individual must have:

- Sufficient knowledge of the specific tasks to be undertaken and the risks which the work will entail.

- Sufficient experience and ability to carry out their duties in relation to the project; to recognise their limitations and take appropriate action in order to prevent harm to those carrying out construction work, or those affected by the work.

- All people who have duties under CDM 2007 should:

  ○ take 'reasonable steps' to ensure persons who are appointed are competent

  ○ not arrange for, or instruct, a worker to carry out or manage design or construction work unless the worker is competent

  ○ not accept an appointment unless they are competent.

Note that:

- The regulations apply to corporate and individual competence.

- Assessment should focus on the needs of the particular project and be proportionate to the risk, size and complexity of the work.

- CDM 2007 should streamline the competence assessment process.

- A key duty of the CDM co-ordinator is to advise the client about the competence of those employed by the client.

- **Worker engagement** is the participation by workers in decisions made by those in control of construction activities, in order that risk on site can be managed the most effective ways.

- **Communication** of the right information, to the right people at the right time to enable them to make appropriate decisions on health and safety issues relating to construction projects.

- The key CDM 2007 Regulations on worker engagement and consultation are:

  ○ Reg. 5: co-operation and Reg. 6 co-ordination

  ○ Reg. 10: client to provide information

    ○   Reg. 11: designers to provide information

    ○   Reg. 13: contractors to provide information [including site induction, information on risks, site rules, imminent danger procedures, training as required by Reg. 13(2)(b) MHSWR].

■   Provide employees with training when exposed to new/increased risks due to being transferred (new site), change of responsibilities, new equipment, new technology, new system of work, etc.

■   For all projects, dutyholders should:

    ○   provide information needed to carry out work without risk

    ○   provide site-specific induction

    ○   advise on findings from risk assessment

    ○   explain site rules

    ○   explain what to do if imminent danger

    ○   advise who is responsible for implementing health and safety on site.

■   Allow for those who either cannot read or may not understand English.

■   Workers have the duty to report anything that will endanger themselves or others.

■   Worker safety representatives are entitled to employer-funded training.

Arrangements for serious danger under Reg. 8 MHSWR:

■   communicate to workers what to do for a notifiable project the principal contractor must:

    ○   make and maintain arrangements to ensure co-operation and consultation between themselves, contractors and workers

    ○   to carry out consultation with workers.

CDM 2007: further advice

Further advice about the CDM 2007 Regulations is obtainable from:

■   CDM 2007 Regulations and Approved Code of Practice (L144)

■   HSE website: www.hse.gov.uk/construction/cdm.htm

■   CDM 2007 industry guidance:
www.cskills.org/supportbusiness/healthsafety/CDMRegs.

Design issues:

■   www.dbp.org.uk

■   www.dqi.org.uk

■   www.cic.org.uk

■   www.ciria.org.uk.

APPENDIX 3 Project planning

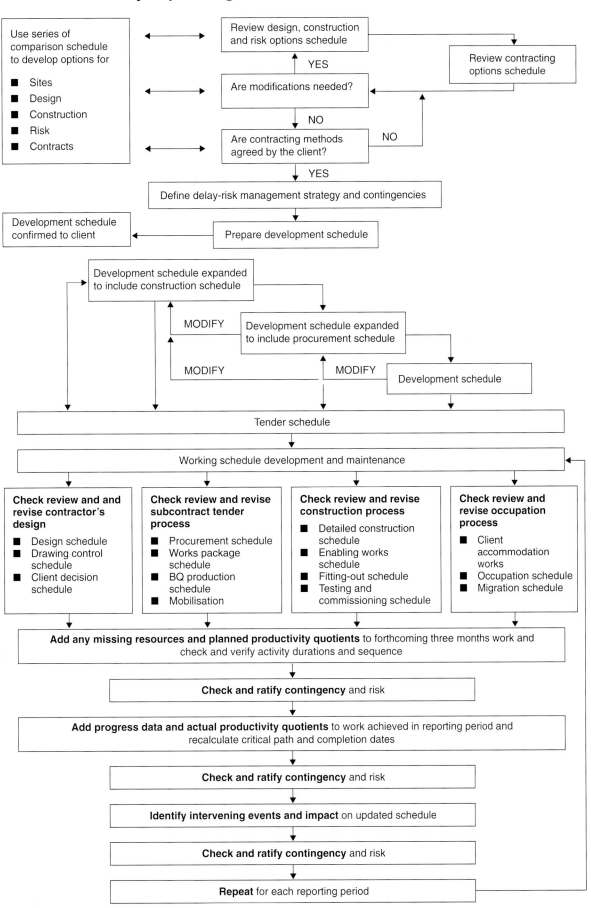

APPENDIX 4   Site investigation

This flow chart is used for each of the 10 activities identified in the table below.

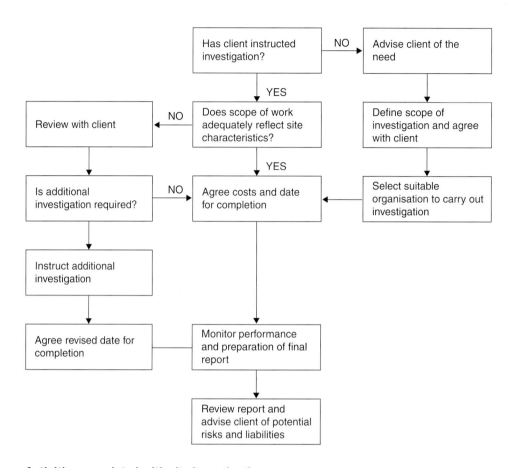

Activities associated with site investigation

Activity	Action by
Site surveys	Land surveyor and structural engineer
Geotechnical investigation	Ground investigation specialist
Drainage and utilities survey	Civil engineering consultant
Contamination survey	Environmental and/or soil specialist
Traffic study	Transportation consultant
Adjacent property survey	Buildings/party walls/rights of light surveyors
Archaeological survey	Local museum or British Museum and other relevant sources
Sustainability issues	Specialist consultant
Legal aspects	Solicitor
Outline planning permission	Architect

Confirmation that the activities have been successfully completed is the responsibility of the project manager.

Each task can be broken down into a number of specific elements.

Site surveys
- location
- Ordnance Survey reference

- ground levels/contours
- physical features (e.g. roads, railways, rivers, ditches, trees, pylons, buildings, old foundations, erosion)
- existing boundaries
- adjacent properties
- site access
- structural survey
- previous use of site,

Geotechnical investigation
- trial pits
- boreholes and borehole logs
- geology of site including underground workings
- laboratory soil tests
- site tests
- groundwater observation and pumping tests
- geophysical survey.

Drainage and utilities survey
- existing site drainage (open ditch, culvert or piped system)
- extent of existing utilities on or nearest to the site (water, gas, electricity, telecoms)
- extent of any other services that may cross the site (e.g. telephone/data lines, oil/fuel pipelines).

Contamination survey
- asbestos
- methane
- toxic waste
- chemical tests
- radioactive substances.

Traffic survey
- examination of traffic records from local authority
- traffic counts
- traffic patterns
- computer simulation of existing traffic flows
- delay analysis
- noise levels.

Adjacent property survey

Traffic survey
- right of light
- party-wall agreements
- schedule of conditions
- foundations
- drainage

- access
- public utilities serving the property
- noise levels (e.g. airports, motorways, air-conditioning equipment).

Archaeological survey
- examination of records
- archaeological remains.

Sustainability issues
- effects of proposed development on local environment
- environmental impact assessment
- flood risk
- carbon dioxide emissions
- waste
- transport
- pollution
- ecology and biodiversity
- health and well-being
- social issues.

Legal aspects
- ownership of site
- restrictive covenants
- easements, e.g. rights of way, rights of light
- way-leaves
- boundaries
- party-wall agreements
- highways agreements
- local authority agreements
- air rights.

Outline planning permission
- effect of local area plan.

APPENDIX 5 Guidance on EU procurement rules

The European Union (EU) Procurement Directives, and the regulations that implement them in the UK, set out the law on public procurement. Their purpose is to open up the public procurement market and to ensure the free movement of goods and services within the EU.

The rules apply to purchases by public bodies and certain utilities which are above set monetary thresholds. They cover all EU Member States and, because of international agreements, their benefits extend to a number of other countries.

Where the regulations apply, contracts must be advertised in the *Official Journal of the European Union* (OJEU) (unless it qualifies for a specific exclusion, e.g. on grounds of national security) and there are other detailed rules that must be followed. The rules are enforced through Member States' courts, and the European Court of Justice (ECJ).

What are the key changes?

The changes introduced through the current set of regulations (enacted on 31 January 2006) include:

- Supply, services and works are consolidated into a single set of regulations.

- Framework agreements and e-auctions expressly included.

- A new competitive dialogue procedure introduced in addition to the open and restricted procedures.

- A dynamic purchasing system introduced.

- Specific provisions made for central purchasing bodies.

- Mandatory exclusion of entities whose directors or other decision-makers have been convicted of certain offences.

- A 10-day standstill period at the award stage prior to contract signature has been provided for.

What about mixed contracts?

- Where a contract covers both services and supplies, the classification should be determined by the respective values of the two elements.

- Where a contract covers works/supplies or works/services, it should be classified according to its predominant purpose.

- Where a contract provides for the supply of equipment and an operator it should be regarded as a services contract.

- Contracts for software are considered to be for supplies unless they have to be tailored to the purchaser's specification, in which case they are services.

What is the advertisement requirement?

Generally, contracts covered by the regulations must be the subject of a call for competition by publishing a contract notice in the OJEU. In most cases the time allowed for responses or tenders must be no less than a set period, although some reduction is possible under certain circumstances (see SIMAP website for further details).

There are some services (categorised as Part A and Part B services) where a reduced advertisement requirement applies: details of this are available on SIMAP website (http://simap.europa.eu).

Below is a table outlining the advertisement timescale requirements.

Procedure	Text	Days
Open	Minimum time for receipt of tenders from date contract notice sent. Reduced when prior information notice (PIN) published (subject to restrictions) to, generally, 36 days and no less than 22 days	52
Restricted	Minimum time for receipt of requests to participate from the date contract notice sent	37
	Minimum time for receipt of tenders from the date invitation sent. Reduced when PIN published (subject to restrictions) to, generally, 36 days and no less than 22 days	40
Restricted Accelerated	Minimum time for receipt of requests to participate from the date contract notice sent	15
	Minimum time for receipt of tenders from the date invitation sent	10
Competitive dialogue and competitive negotiated	Minimum time for receipt of requests to participate from the date contract notice sent	37
Competitive negotiated accelerated	Minimum time for receipt of requests to participate from the date contract notice sent	15

What are the procurement options?

- Open procedure: all interested parties can respond.

- Restricted procedure: a selected number of respondents are invited to tender.

- Competitive dialogue procedure: following an OJEU contract notice and a selection process, the authority then enters into dialogue with potential bidders, to develop one or more suitable solutions for its requirements and on which chosen bidders will be invited to tender.

- Negotiated procedure: a purchaser may select one or more potential bidders with whom to negotiate the terms of the contract. An advertisement in the OJEU is usually required but, in certain circumstances, described in the regulations, the contract does not have to be advertised in the OJEU. An example is when, for technical or artistic reasons or because of the protection of exclusive rights, the contract can only be carried out by a particular bidder.

What is the impact of the regulations on private sector projects?

For public works concession contracts (i.e. contracts under which the contractor is given the right to exploit the works, e.g. tolled river crossings), the winning concessionaire is required to comply with certain OJEU advertising requirements for works contracts which it intends to award to third parties. For some subsidised works contracts (civil engineering activities, building work for hospitals, facilities intended for sports, recreation and leisure, school and university building or buildings for administrative purposes) the public authority awarding the grant is obliged to require the subsidised body to comply with the regulations, as if it were a public authority, as a condition of grant. This provision has, for example, been invoked for many lottery-funded projects. There is a similar requirement for subsidised service contracts in connection with subsidised works.

Endnote

This guidance is not intended as a substitute for project-specific legal advice, which should always be sought by a public authority where required. The EU procurement regime is not static. It is subject to change, driven by evolving European and domestic case law, European Commission communications, new and revised Directives and amendments of the existing UK regulations. Further information can be obtained from SIMAP or the Office of Government Commerce.

Part 1 Appendices

APPENDIX 6 Performance management plan

Performance management should be an integrated part of a development project from its definition through to monitoring and review. Where it is not possible to establish direct 'cause and affect' linkages or precise measures of performance, interim measures such as key performance indicators (KPIs) are often used. These could be trends over time, value to the customer, awareness of product or service.

Objectives

The purpose of a performance management plan (PMP) is to set out the principles and targets for a schedule against which it delivers its outputs, outcomes and benefits. The plan also defines how the performance criteria will be measured and plans for any divergence management. The plan contains details of the performance management process, performance measurements and the performance information required to establish and monitor delivery.

Performance management process

The performance management process outlines the activities to set direction, which uses performance information to manage better, demonstrates what has been accomplished and sets actions to improve. Performance metrics may be defined using the SMART test (Specific, Measurable, Attainable, Relevant and Timely).

Performance measurements should indicate milestones for measuring progress against goals (may be at the key decision stages), against target levels of intended accomplishment (target objectives) and against third parties. Measures may need to change as progress is made. Measurement criteria may be defined using the FABRIC test (Focused, Appropriate, Balanced, Robust, Integrated, and Cost-effective).

Performance information includes the data, their characteristics, quality, sources and contribution to a measure.

Checklist for PMP

Quality criteria for a PMP include:

- Are the objectives, outputs, outcomes or benefits against which to set and monitor performance or achievement of targets clearly defined?

- Can the performance measures be assessed against key objectives?

- Are the performance measures clearly defined, together with target values?

- Is the approach for managing performance is complete and does it contain all key elements as a cycle of activities?

- Do the measures and metrics criteria meet pre-defined tests of SMART and FABRIC (if applicable)?

- Is the periodicity of measurement clearly defined?

- Have all standards or techniques to be used for measurement are defined?

- Are the sources of performance information of adequate quality?

- Is the proposed performance information reliable and/or independently validated?

- Is the approach to investigation and corrective action to improve unsatisfactory performance clearly defined?

- Is there an outline of the management organisation and process?

- Are the resources to collect and analyse performance information clearly defined?

- Are the roles and responsibilities clearly defined?

Suggested contents for the PMP

The key elements of performance management plan will describe a cycle of activities and their outputs:

- Strategy: defining the aims and objectives of the organisation.

- Selection of performance measures: identifying the measures which support the quantification of activities over time.

- Selection of targets: quantifying the objectives set by management, to be attained at a future date.

- Delivery of performance information: providing a good picture of whether an organisation is achieving its objectives.

- Reporting information: providing the basis for internal management monitoring and decision-making, and the means by which external accountability is achieved.

- Action to improve: taking action to put things right; feeding back achievements into the overall strategy of the organisation.

Part 1 Appendices

APPENDIX 7 Implications of the Housing Grants, Construction and Regeneration Act 1996, Amended 2009

The Housing Grants, Construction and Regeneration Act 1996 is applicable to all construction contracts entered into after 1 May 1998. The intention behind introducing the reforms pinned around improving cash flow, reducing confrontation and facilitating 'fair play' in the payment mechanisms. In order to enhance the effect of this, a number of amendments (initially proposed in 2004) have been introduced from September 2009 focusing on increased transparency and clarity, encouraging parties to resolve disputes by adjudication and improving the right to suspend performance under the contractual arrangements.

The Act is applicable to all construction operations including site clearance, labour only, demolition, repair works and landscaping. However, off-site manufacturing, supply and repair or plants in process industries, domestic construction contracts, contracts not in writing and certain other activities including PFI contracts (but not the construction contracts entered into as a result of PFI) have been excluded from the purview of this Act.

The two key areas affected by this Act are the payment procedures and the adjudication in case of disputes.

Payment under the Act

The Act requires that every construction contract must contain the following elements:

■ payment by instalments

■ adequate mechanism to determine what amounts of payments are due and when

■ prior notice of amounts due and make up

■ prior notification (seven days) of intention to withhold payment (set-off), giving grounds and amounts

■ suspension of work (not less that seven days' notice) for non-payment of payments due

■ all 'pay when paid' clauses outlawed except in the instance where a third party on whom the payment depends becomes insolvent

■ in the absence of minimum requirements as specified by this Act, the government scheme comes into operation as a default clause.

The government scheme for payment

The government scheme for payment provides for:

■ monthly interim payment

■ due date for interim payments is seven days after the end of the relevant monthly period or from making a claim, whichever is the later

■ final due date for interim payment is 17 days from due date

■ notice of amount due should be given not later that five days after due date

■ notice of intention to withhold payment should be given not later than seven days before final payment date.

Adjudication under the Act

The Act makes it a statutory right to refer any dispute for adjudication. It stipulates that all contracts must contain an adjudication procedure that complies with the Act:

■ Either party can give notice of adjudication at any time regarding any dispute or difference arising under the contract.

■ The contract must provide a timetable for appointment of adjudicator and referral of dispute within seven days of initial notice.

■ The adjudicator must reach decision within 28 days of referral (up to 42 days if the referring party agrees).

■ The adjudication period can be extended only if parties agree, or on the adjudicator's instigation with the consent of the referring party.

■ The adjudicator is enabled to take necessary initiatives in ascertaining facts and law.

■ The decision of the adjudicator is binding until the dispute is finally determined by legal proceedings or arbitration or by agreement.

■ The parties have the option to agree to accept the adjudicator's decision as final.

■ The government scheme comes into operation as a default mechanism if the minimum requirements as stipulated by the Act are not met.

The government scheme for adjudication

The government scheme for adjudication provides for:

■ Written notice of adjudication stating:

    ○ nature and description of dispute and parties involved

    ○ details of where and when dispute has arisen

    ○ nature of redress sought

    ○ names and addresses of parties to contract.

■ The appointment of an adjudicator within seven days of notice.

■ The same seven days in which to submit full documentation (referral notice).

■ Oral evidence limited to one representative (may or may not be a lawyer).

■ The adjudicator's decision within 28 days from referral notice or 42 days with the referring party's permission.

■ Parties are equally responsible for payment of the adjudicator's fees (unless otherwise determined by the adjudicator).

■ Reasons for the adjudicator's award have to be provided if requested.

■ The decision by the adjudicator is binding pending any final determination by legal proceedings or arbitration, or by mutual agreement in settlement.

■ Parties have to comply with the adjudicator's decision immediately.

Part 1 Appendices

A overview of adjudication

Adjudication is intended to be quicker and less expensive than court proceedings. Therefore the parties must be prepared for a degree of 'rough justice'. The adjudicator has very wide powers. He can use his initiative and can request further documents from any party, meet and question them, visit the site, appoint experts to help him if necessary (e.g. technical assessors, legal advisors, etc.), issue directions and timescales. He can adjudicate, with the consent of all parties, on 'related disputes' under different contracts. He can award interest payments.

What can be referred to the adjudicator?

Virtually all kinds of dispute or difference may be referred to the adjudicator provided they arise 'under the contract' (the contract can be written, oral or partly oral). These would include: failure to issue notice of sums due or notice of withholding payment, value of interim payments, value of variations, extension of time, loss and expense, set-off and contra-charges, workmanship, whether or not an instruction was reasonable, etc.

Who pays for the cost of adjudication?

It is usual that each of the parties must bear their own costs in submitting and presenting their cases. However, under some circumstances a 'costs paid by loser' approach may be undertaken.

How is the adjudicator's decision enforced?

The adjudicator's decision is intended to be binding pending final determination by legal proceedings or arbitration, or by mutual agreement in settlement. Several court cases have shown that the courts intend to support both the Act and the adjudicator by enforcing awards.

It may well be that the mere presence of an adjudication resource will concentrate the minds of those on either side whose stance is less than reasonable and so enable the parties to go forward with providing the client's end product – the completed project – on time and free of major disputes.

APPENDIX 8   Guidance on partnering

What is partnering?

Partnering is a management approach used by two or more organisations to achieve specific business objectives by maximising the effectiveness of each participant's resources. It requires that the parties work together in an open and trusting relationship based on mutual objectives, an agreed decision-making process and an active search for continuous measurable improvements.

Partnering is the most efficient way of undertaking all kinds of construction work including new buildings and infrastructure, alterations, refurbishment and maintenance. The basic elements have been in practice for a long time, but the benefits have not been recognised by those involved. The elements are evident when those involved on a project have worked together before and are brought into the project at the very early stages. In this ideal situation everyone naturally works together as a team.

Partnering can be based on a single project, but the real benefits are realised when it is based on a long-term strategic commitment. Project-specific partnering is about partnering on individual projects. Strategic partnering is about long-term relationships between parties who are prepared to work together over extended periods of time. By building on the individual strengths of the separate businesses, a strategic partnering arrangement can deliver steadily improved performance over several years.

Definition of project partnering

Project partnering is a set of actions taken by the work teams that form a project team to help them co-operate in improving their joint performance. Specific actions are agreed by the project team, taking account of the project's key characteristics, and its own experience and normal performance. The choice of actions is guided by a structured discussion of mutual objectives, decision-making processes, performance improvements and feedback.

Project partnering involves initial costs and provides substantial benefits. It is not a fixed way of working; it develops as project teams co-operate in finding the most effective ways of achieving agreed objectives.

Definition of strategic collaborative working

Strategic collaborative working is a set of actions taken by a group of firms to help them co-operate in improving their joint performance over a series of projects.

The actions initially aim to agree an overall strategy, ensure the right firms are included, financial arrangements support partnering, firms' cultures, processes and systems are integrated, overall performance is benchmarked, project processes are continually improved and the whole strategic partnering arrangement is guided by feedback. Ultimately, the actions aim at establishing and continuously developing a long-term business based on an integrated construction cycle that links clients' use of constructed facilities with their development and production.

The more partnering is used and the benefits are experienced, the more likely it is that strategic long-term partnering will be used. Partnering is about the formation and development of a relationship or relationships which benefit construction projects. Partnering creates a variety of opportunities and concerns for the participants. These include early involvement of suppliers, selection of all parties by value, performance measurement and continuous improvement, common team processes and commercial arrangements that align risk and reward for all parties on both the demand and supply sides of the industry. The term 'collaborative working' is often used to help eliminate misconceptions about previous definitions of partnering. This appears to encourage people to invest in learning about new best practice methods and adopt co-operative behaviour.

Essential features of partnering

Partnering consists of three essential features: mutual objectives, an agreed decision-making process and an active search for continuous improvement. They form the partnering logo shown below:

Essential actions of project partnering

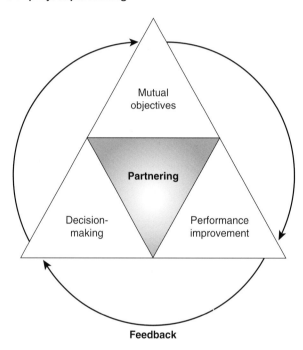

Mutual objectives

The most fundamental requirement for partnering is to agree mutual objectives. The aim is to find objectives that firmly establish for everyone involved in the project that their own best interests will be served by concentrating on the overall success of the project. When people co-operate in adopting a 'win–win' attitude they increase the chances that they will produce enough for everyone to have everything they reasonably want.

Partnering accepts that firms look after their own interests. It requires a tough-minded recognition by clients that they will get what they need only if consultants, contractors and specialists have a realistic opportunity to do good work and make reasonable profits. It requires an equally tough-minded recognition by consultants, contractors and specialists that they prosper best when clients get excellent value, good buildings or infrastructure and no hassle. This focus on mutual objectives gives expression to the idea that when people co-operate they can produce more than enough to give everyone what they reasonably want. This is often described as a 'win–win' attitude in contrast to the traditional zero-sum assumption that if one person gains, someone else must lose.

Clients should ensure that agreed mutual objectives take account of the interests of everyone affected by the project. It may take time to deal with everyone's concerns. Inevitably the client, designers, managers, specialist contractors and manufacturers have different views about what constitutes success. And people tend to worry that in some way they will lose out if they co-operate in meeting other peoples' needs. Despite these perhaps natural reservations, experience shows that when project teams are brought together to discuss their individual interests they can find mutual objectives. The initial costs and time involved are worthwhile because well thought out mutual objectives avoid time and resources being wasted on designs likely to create problems at later stages.

Mutual objectives may deal with many process and product related issues but common subjects include:

- value for money
- guaranteed profits
- reliable quality
- fast construction
- handover to owner on time
- cost reduction
- costs within agreed budget
- operating and maintenance efficiency
- improved efficiency for users
- architectural quality
- a specific technical innovation
- excellent site facilities
- safe construction
- shared risks
- timely design information
- reliable flow of design information
- shared use of computer systems
- effective meetings
- training in decision making skills
- training in management control systems
- no claims.

In agreeing mutual objectives it is essential to sort out the financial arrangements so that everyone gets a fair return in business terms. The worst situations in construction projects arise when any of the consultants or contractors is losing money. Contractual arrangements should ensure that none of the firms, if they contribute their best efforts, would lose out relative to the others. The essential equity of good value for clients and fair profits for consultants, contractors and specialists provides the platform on which partnering flourishes.

Decision-making process

Construction projects bring together many work teams drawn from many different firms. These work teams need to agree how decisions will be made. The nature of decision-making systems is directly influenced by whether the client needs the project to produce an existing well-developed answer or an original design. An important consequence of this choice is the amount of time the client and his staff will need to spend in making decisions. Original designs take more time but should result in buildings and infrastructure that support the users' activities and delight everyone who sees them. Standard answers make fewer calls on the client's time, are quicker, cost less and provide reliable quality but may well impose more compromises on users and look dull.

The project team need to agree the information and communication systems they will use. They also need to decide the quality, time and cost-control systems they will use. They need to agree who will operate these systems and who will get the various outputs. They decide on the form and frequency of face-to-face meetings. They consider the use of taskforces, workshops, common project offices, social events and other ways of bringing teams closer together. Overall, the systems should ensure that good ideas are captured and properly considered. This means balancing the benefits of using an existing satisfactory answer, and the greater benefits, costs and time of finding and developing something better.

Whatever decision-making systems are agreed they should include robust procedures to ensure that problems are resolved quickly in ways that encourage co-operative teamwork. This means most problems are resolved by the work teams directly involved. When a problem cannot be resolved in this way, it should be referred immediately to the project's core team and in exceptional cases to senior managers.

Continuous performance improvement

The main aim of partnering is to improve project teams' performance. Partnering that merely provides mutual objectives and agreed ways of making decisions will drift into inefficient ways of working. Partnering requires project teams to search for better answers. Project teams new to partnering should aim at one modest improvement that all members of the team regard as important. As experience of partnering grows, the scale and range of improvements will increase.

It is important that performance improvements in one area do not distract work teams from continuing to deliver their established normal performance in all other areas. This is an easy trap to fall into as attention is focused on the improvements and quality elsewhere slips without anyone noticing. This is why partnering procedures give explicit attention to the constraints of achieving normal performance as well as delivering performance improvements.

There is much discussion about the best ways of encouraging work teams to improve their performance. Many economists argue that competition provides the best incentives to improve performance. However, competition in the construction industry can easily become cut-throat and bid prices, quality and safety are driven down to levels that are hopelessly inefficient. The outcomes include claims, disputes, defects, late completions and good firms being driven out of business. Competition has a place in partnering to encourage consultants, contractors and specialists to invest in training and innovation to improve their own performance. This can be achieved even when there are long-term relationships between firms.

By having two, three or four options available for key relationships, all the partners are motivated to continuously improve their performance.

There is usually provision in partnering contracts for incentives to be agreed, which may allow for participants to share in cost savings achieved, but may also provide for a similar distribution of losses due to errors or cost increases. These arrangements are known as gain share/pain share.

Benchmarking provides another weapon in the search for improved performance. Carefully researched information about best international practice is often used by experienced clients to guide the choice of targets. A good approach is to concentrate on whatever the client, consultants or contractors regard as their biggest problem.

There are advantages in project teams setting their own targets. When teams are given good information about the performance achieved by leading practice, they often set tougher targets than any they would accept from their managers.

Having agreed the performance improvements they will aim for, the best partnering teams try various ideas; continue with actions that work and change those that deliver no improvements. They often set up a task force to help find ways of meeting targets. This is a small group of people with relevant knowledge selected from within the project team and it may include external experts. Task forces should be given a short time to find an innovative answer that will deliver significant performance improvements.

The first partnering workshop should establish procedures to ensure that innovations and new actions found to deliver improvements will be built into standards and procedures for the benefit of the current and future projects.

Feedback

Teams need to be guided by feedback about their own performance if they are to deliver the substantial benefits that partnering can provide. Achieving performance improvements depends on project teams being provided with up-to-date and objectively measured feedback. Teams should measure their own performance and plot the results on control charts that show graphically how they are doing against their targets. Teams believe feedback they have produced themselves and use it to search for better ways of working. Feedback is most effective when it is expressed in positive terms. For example, quality should be measured by recording how often quality standards are achieved, not the number of failures.

Performance improves faster when successes are publicised and celebrated. It is vital that senior managers know when targets are being achieved and make a point of congratulating and rewarding the people involved. The rewards can be token, but a dozen cans of lager presented at a light-hearted ceremony to the week's best work team can ensure that all the teams strive to be winners next week.

Failures must not be ignored. This is not to allocate blame which is counterproductive. Failures should be used to guide teams in looking for robust answers to problems so that performance is back on target quickly. Some effective teams make a point of celebrating failures because they provide opportunities to find more effective ways of working. When a failure arises, they have a party and then, with renewed enthusiasm, concentrate on finding a robust answer. It is important that senior managers are kept up to date about improvements in performance. This is essential if they are to remain committed to partnering. At least some managers in most organisations take a pride in being highly competitive and are sceptical of the idea that the co-operative methods used in partnering can possibly be effective. Without regular, well-founded feedback on the performance improvements delivered by partnering, there is always a risk that adversarial methods will be reintroduced.

Feedback should flow from project to project. Too many innovative ideas are lost because of weak feedback systems. Lessons need to be captured so that good ideas are applied on future projects and problems and defects do not recur. Leading firms involved in partnering have developed standards and procedures that systematically capture best practice as it emerges from their projects. The feedback-based standards and procedures help all their project teams concentrate on efficient work. This is an essential element in using strategic partnering and strategic collaborative working successfully.

Maintaining partnering throughout projects

The first partnering workshop on any project is important in giving the project team a firm basis for ensuring that partnering delivers benefits. However, best practice includes workshops throughout projects to review progress and if necessary change things agreed at the first partnering workshop. Change may be in response to the project going better than expected and the team realising they can aim for bigger performance improvements. It is perhaps more common for projects to face problems. These should be discussed at a workshop, which if the problem is sufficiently serious, should be called especially. The workshop should look for and agree actions that deal with persistent problems once and for all. Partnering is action oriented and dealing with problems quickly is central to its success.

A final workshop is used to identify good ideas and lessons identified during the project so they can be recorded and made available for use on future projects.

Partnering is an ongoing activity guided by workshops which all need to be taken seriously especially by the senior managers involved. The potential benefits are large and they are earned by concentrating on and continually reinforcing co-operative teamwork. CIOB's *Partnering in the Construction Industry: A Code of Practice for Strategic Collaborative Working* provides further detailed guidance and advice.

APPENDIX 9 Project risk assessment

Risk is inherent in almost all construction projects. Depending on the nature and potential consequences of the uncertainties, measures are taken to tackle risks in various ways. Risk management is a systematic approach to identifying, measuring, analysing, and controlling areas or events with a potential for causing unwanted change. It is through risk management that risks to the schedule and/or project are assessed even though some risks are speculative and may bring about gains as well as losses.

A process of risk assessment and management has to be implemented at an early enough stage to have an impact on the decision-making during the development of the project. Small workshops should put together a list of events, which could occur, that threaten the assumptions of the project. A typical list requiring different workshop forum might be:

- revenue

- planning consent

- schedule

- design

- procurement and construction

- maintenance and operation.

Assessments should be made on:

- probability of occurrence (percentage)

- impact on cost–time–function (money–weeks–other)

- mitigation measures

- person responsible for managing the risk

- delete point where the risk will have passed (date)

- date for action by – transference – insurance – mitigation (date).

The mitigation measures need evaluating in terms of good value for money. Transference should be to the party most capable of controlling it since they will commercially price it the lowest. Lateral thinking might shrink, or in rare cases eliminate, risk. The risk register should be regularly reviewed and be part of all onward decision-making processes.

Risk register

The formal record for risk identification, assessment and control actions is the risk register. The risk register may be divided into three parts as follows:

- Generic risks: risks that are inherent irrespective of the type or nature of the project.

- Specific risks: those risks that are related to the particular project, perhaps identified through a risk workshop involving the project team.

- Residual risks: this is a list of risks identified above which cannot be excluded or avoided and contingency has to be provided for their mitigation.

The financial effect of such residual risks must be evaluated to determine an appropriate contingency allowance. A time contingency should also be considered.

When using the risk registers, values of occurrence and consequence are assessed using high, medium and low (H, M, L) values. Other forms of assessments include very high, high, medium, low and very low scoring or using numerical values (e.g. a ranking of 1–10, with 10 signifying a very high probability/impact risk and 1 identifying a risk of almost negligible impact/probability). An example of a part of a risk register is shown below.

Risk number	Description	Probability of occurrence %	Unmitigated impact			Mitigations adapted	Person responsible	Last updated	Delete point	Date to action
			Impact cost	Time	Function					
1										
1.1										
1.2										
1.3										
2										
2.1										
2.2										
2.3										
2.4										
2.5										
2.6										
2.7										
2.8										
2.9										
2.10										
2.11										
2.12										
2.13										
2.14										
2.15										
2.16										
2.17										
2.18										

Often risks are external arising from events one cannot influence. This particularly applies to markets and revenue projections. Some might even be 'show stoppers'. Value for money studies might throw up options of hedging or taking out special insurance. Obviously the risks of high probability and high impact are the ones on which to concentrate.

Contingency planning

Contingency planning is the development of a strategy to minimise the effects of intervening events that could possibly interfere with the smooth running of the project at some time between inception and completion. A contingency is a planned allotment of the time likely to be taken up by the occurrence of an intervening event.

Some scheduling software will offer a choice as to whether activities are, by default, to be scheduled as early as possible or as late as possible. Others default to one or the other. Where an activity would otherwise be scheduled as late as possible, the introduction of a contingency period buffering its end date will have the effect of scheduling the planned commencement of the activity earlier than would otherwise be the case. The effect of this will provide for a degree of delay in the completion of the activity to be absorbed by the contingency period. Designated non-work periods such as religious, industry-related or statutory holidays or weekends are not contingency periods and should not be treated as such. Only that party who is contractually liable for the consequences of the risk maturing can properly determine the quantity and distribution of the contingency it perceives to be required from time to time. Accordingly, contracts should (and generally do) make clear who is contractually liable for the consequences of the risk maturing and accordingly who owns the contingency.

In the same manner that cost budgets usually have an allocation of funding called 'a contingency sum' that the employer may rely on to spend against for unforeseen events, the schedule must have strategically placed contingency activities to absorb the time effect of intervening events that are at the employer's risk.

Prudent contractors will also make allowances for the risks they bear in the management and distribution of the resources and the quality of the work they carry out.

Contingencies should be designed to be identified separately for both the employer's and the contractor's risks and for those risks which are related to:

- ■ an activity, or chain of activities

- ■ a contractor, subcontractor, supplier or other resource

- ■ an access or egress date or date of possession or relinquishment of possession

- ■ the works, any defined section, and any part of the works.

At the lowest level of density, schedule contingencies are likely to be the longest in order to provide some accommodation for the unknown aspects of the schedule. Because of the absence of precision at this level of density, the separately allocated contingencies to one party or the other may both be arrived at by a formula adjustment. One way of identifying contingencies at this level is to use a formulaic approach such as Monte Carlo analysis to allocate an additional period to the known activities. The Monte Carlo algorithm randomly generated values for uncertain variables over and over again to generate model contingency periods.

At medium density, there is little scope in the schedule for notional formulaic calculations to accommodate unknown and unquantifiable risks and contingencies must be clearly allocated to one part or the other. There must be no contingency that is unallocated to an owner. At medium density, the risks should be clearly identified and a rational explanation set down in the method statement of the manner in which the possibility of the risk maturing has been allowed for.

At high density, the risks that need to be accounted for will be significantly fewer than at other densities. At this density contingencies must be clearly allocated to one party or the other. There must be no contingency that is unallocated to an owner and none clearly justified in the method statement. There may legitimately be risks such as inclement or adverse weather, unforeseeable ground conditions and utilities, plant breakdown, re-work, absenteeism which may need to be allowed for at this density but there should not be the need for design risk contingencies, or implied variations at this stage.

An example of a format of a risk mitigation table is shown below.

Mitigation action plan

RISK NAME:	DATE:	ISSUE No.:	ISSUED BY:

Risk category/Ref.:		Risk ownership:	

Risk evaluation		Probability	Cost	Time	(Other area e.g. environment/ health and safety)	Total score
		Current				
	Projected					

Risk description:

Risk mitigation plan:	By whom:	Review point/milestones:
End date/time scale		

Comments

Part 1 Appendices

Project risk assessment checklist

Project _____   Date _____

Overall risk assessment is:

Signatory	Signature	Date
Client		
Project manager		

☐ Normal risk

☐ High risk

Risk consideration		Criteria	Risk assessment		Proposed management of high risk
			Normal	High	
1 Project environment					
User organisation	○	Stable/competent	☐ ☐	☐ ☐	
	○	Poor/unmotivated/untrained			
User management	○	Works as a team	☐ ☐	☐ ☐	
	○	Factions and conflicts			
Joint venture	○	Client's sole contractor	☐ ☐	☐ ☐	
	○	Third party involved			
Public visibility	○	Little or none	☐ ☐	☐ ☐	
	○	Significant and/or sensitive			
Number of project sites	○	2 or less	☐ ☐	☐ ☐	
	○	3 or more			
Impact on local environment	○	High	☐ ☐	☐ ☐	
	○	Low			

Risk consideration		Criteria	Normal	High	Proposed management of high risk
2 Project management					
Executive management involvement	○	Active involvement	☐ ☐	☐ ☐	
	○	Limited participation			
User management experience	○	Strong project experience	☐ ☐	☐ ☐	
	○	Weak project experience			
User management participation	○	Active participation	☐ ☐	☐ ☐	
	○	Limited participation			
Project manager	○	Experienced/full-time	☐ ☐	☐ ☐	
	○	Unqualified/part-time			
Project management techniques	○	Effective techniques used	☐ ☐	☐ ☐	
	○	Ineffective or not applied			
Client's experience of project type	○	Client has prior experience	☐ ☐	☐ ☐	
	○	First for client			

Part 1 Appendices

---

3 Project characteristics

Complexity	○	Reasonably straightforward	□ □ □ □
	○	Pioneering/new areas	
Technology	○	Proven and accepted methods and products	□ □ □ □
	○	Unproven or new	
Impact of failure	○	Minimal	□ □ □ □
	○	Significant	
Degree of organisational change	○	Minimal	□ □ □ □
	○	Significant	
Scope	○	Typical project phase or study	□ □ □ □
	○	Unusual phase or study	
Foundation	○	First phase or continuation	□ □ □ □
	○	Earlier work uncertain	
User acceptance	○	Project has strong support	□ □ □ □
	○	Controversy over project	
Proposed time	○	Reasonable allowance for delay	□ □ □ □
	○	Tight/rapid build-up	
Scheduled completion	○	Flexible with allowances	□ □ □ □
	○	Absolute deadline	
Potential changes	○	Stable industry/client/ application	□ □ □ □
	○	Dynamic industry/client/ application	
Work days (developer)	○	Less than 1000	□ □ □ □
	○	1000 or more	
Cost-benefit analysis	○	Proven methods or not needed	□ □ □ □
	○	Inappropriate approximations/ methods	
Hardware/software capacity estimates	○	None or proven methods	□ □ □ □
	○	Unproven methods/ no contingency	

---

4 Project staffing

User participation	○	Active participation	□ □ □ □
	○	Limited participation	
Project supervision	○	Meets standards	□ □ □ □
	○	Below standards	
Project team	○	Adequate skills/experience	□ □ □ □
	○	Little relevant experience	

5 Project costs

Cost quotation	○	Normal (i.e. time-based)	☐	☐	☐	☐
	○	Fixed price				
Cost estimate basis	○	Detailed plan/proven method	☐	☐	☐	☐
	○	Inadequate plan/method				
Formal contract	○	Non-standard form	☐	☐	☐	☐
	○	Standard form				

6 Other

APPENDIX 10 Guidance on value management

Value management and value engineering

Value management (VM) and value engineering (VE) are techniques concerned with achieving 'value for money'. VE was pioneered by an American, Lawrence Miles, during the Second World War to gain maximum function (or utility) from limited resources. It is a systematic team based approach to securing maximum value for money where:

$$Value = \frac{Function}{Cost}$$

Thus, value can be increased by improved function or reduced cost. The technique involves identification of high-cost elements, determination of their function and critical examination of whether the function is needed and/or being achieved at lowest cost. In terms of projects, VE has the greatest influence and impact at the strategy/design stage. It requires reliable and appropriate cost data and uses brainstorming workshops by a group of experts under the direction of a facilitator.

VM is similar to VE, but in terms of projects it focuses on the overall objectives and is most appropriate at the option identification and selection stage where the scope for maximising value is greatest.

Brainstorming forums involving those who would naturally contribute to the project and/or those with a significant interest in the outcome are a fundamental component of both VE and VM. Participants should be free to put up ideas and as far as possible idea generation and analysis should be kept separate. It is the role of the chairman to ensure that this is the case, and hence good facilitation skills and a measure of independence are essential characteristics of the role.

The process

Value techniques are founded on three principal themes:

■ Achievement of tasks through involvement and teamwork; based on the premise that a team will almost always perform better than an individual.

■ Using subjective judgement, which may or may not incorporate risk assessment.

■ Value is a function of cost and utility in its broadest sense.

Key decisions in the application of value techniques are:

■ When should the technique be utilised?

■ Who should be involved?

■ Who should perform the role of the facilitator?

A balance must be struck between early application before an adequate understanding of the problem and constraints has been achieved and late application when conclusions have been drawn and opinions hardened. Although feasibility (when identifying suitable options) and pre-construction (before design freeze) would in most cases be suitable , each project should be examined on its merits.

The facilitator's role is to gain commitment and motivate participants, draw out all views and ensure a fair hearing, select champions to take forward ideas generated and to keep to the agenda. To achieve these goals the facilitator must be independent, possess well-developed interpersonal and communication skills and be able to empathise with all participants. Although he or she must understand the nature of the project this need not be at a detailed level. Large, complex or otherwise difficult projects may warrant employment of an external specialist facilitator. The facilitator's role is crucial to the success of the exercise and care needs to be taken over selection.

The purpose and the agenda for the value forum should be determined by the project team. A value statement should also be produced giving a definition of value in relation to the particular project. For example, value may not be related solely to cost but may also encompass risk, environmental impact, occupational utility, etc. Although the significance of these factors will be project specific, the project team must ensure that this statement reflects corporate policy. The statement is not intended to be a constraint but is used as a benchmark throughout the forum to maintain focus.

Two to three days are usually required for each value forum. The project manager must ensure that all supporting information is available to the forum in summarised form and that expert advice is readily available. The project manager must therefore ensure that personnel with a detailed knowledge of the project participate in the forum.

Link to risk assessment

Value techniques may be used in conjunction with risk assessment where there is a variety of means of managing risk and choices have to be made. The process is particularly valuable in identifying the optimum mitigation approach where risk management options impinge on a variety of project objectives.

In this case risk management objectives are determined, in open forum, alongside overall project objectives. Risk management options are then ranked against the full range of objectives to determine the best option overall.

Potential pitfalls

- Cost (monetary and time) of value meetings can be high.

- At the feasibility/option identification stage, aspects of the technique can conflict with the principles of economic appraisal which seeks to identify an optimum solution by reference to an absolute measure of benefit as opposed to the subjective criteria used in value techniques, e.g. a standard of protection could not be specified as a value objective. If benefits or costs could be assigned to all criteria, there would be no role for VM analysis at the feasibility stage, although the facilitative and team-building aspects of the technique would still be useful.

- Where economic appraisal overrides VM, team-building benefits will be undermined.

- Client participation in the value process could prove to be detrimental in the event of dispute with a consultant or a contractor. However, client participation is an integral part of the process, particularly at the feasibility stage, and must therefore be performed by experienced and knowledgeable staff aware of the contractual pitfalls.

Value techniques should be applied where there is a reasonable prospect of cost saving or substantial risk reduction or where consensus is necessary and difficult to achieve. Examples may include high value or complex projects impinging on a variety of interests and projects where social, economic, environmental and/or intangible benefits are significant but difficult to quantify. Value techniques may be used whenever there is a need to define objectives and find solutions.

An example of utilisation of VM at key stages in a project framework is shown below.

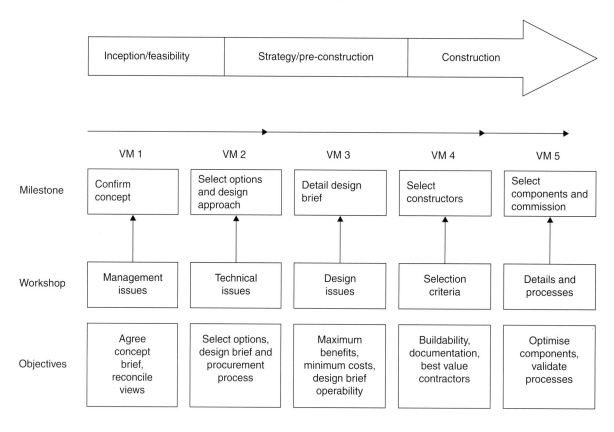

APPENDIX 11 Guidance on environmental impact assessment

Introduction

Environmental impact assessment (EIA) is a key instrument of EU environmental policy. Since passage of the first EIA directive in 1985 (Directive 85/337/EEC) both the law and the practice of EIA have evolved. An amending directive was published in 1997 (Directive 97/11/EC), further amended in 2003 (Directive 2003/35/EC).

Assessment of the effects of certain public and private projects on the environment is required under the Town and Country Planning impact (Environmental impact Assessment) (England and Wales) Regulations 1999 (amended 2008) SI, in so far as it applies to development under the Town and Country Planning Act 1990. EIA is a means of drawing together, in a systematic way, an assessment of a project's likely significant environmental effects. This helps to ensure that the importance of the predicated effects, and the scope for reducing them, are properly understood by the public and the relevant competent authority before it makes its decision.

Where an EIA is required there are three broad stages to the procedure:

■ The developer must compile detailed information about the likely main environmental effects. To help the developer, public authorities must make available any relevant environmental information in their possession. The developer can also ask the 'competent authority' for its opinion on what information needs to be included. The information finally compiled by the developer is known as an environmental statement (ES).

■ The ES (and the application to which it relates) must be publicised. Public authorities with relevant environmental responsibilities and the public must be given an opportunity to give their views about the project and ES.

■ The ES, together with any other information, comments and representations made on it, must be taken into account by the competent authority in deciding whether or not to give consent for the development. The public must be informed of the decision and the main reason for it.

The regulations

The regulations integrate the EIA procedures into the existing framework of local authority control. These procedures provide a more systematic method of assessing the environmental implications of developments that are likely to have significant effects. EIA is not discretionary. If significant effects on the environment are likely, EIA is required. Where the EIA procedure reveals that a project will have an adverse impact on the environment, it does not follow that planning permissions must be refused. It remains the task of the local planning authority to judge each planning application on its merits within the context of the development plan, taking account of all materials considerations, including the environmental impacts.

For developers, EIA can help to identify the likely effects of a particular project at an early stage. This can produce improvements in the planning and design of the development, in decision-making by both parties, and in consultation and responses thereto, particularly if combined with early consultations with the local planning authority and other interested bodies during the preparatory stages. In addition, developers may find EIA a useful tool for considering alternative approaches to a development. This can result in a final proposal that is more environmentally acceptable, and can form the basis for a more robust application for planning permission. The presentation of environmental information in a more systematic way may also simplify the local planning authority's task of appraising the application and drawing up appropriate planning conditions, enabling swifter decisions to be reached.

For EIA applications the period after which an appeal against non-determination may be made is extended to 16 weeks.

Environmental impact assessment (EU regulations)

EIA is a procedure required under the terms of EU directives on assessment of the effects of certain public and private projects on the environment.

Key stages	Notes
Project preparation	The client prepares the proposals for the project.
Notification to competent authority	The competent authority may be the Environment Agency, English Nature or a similar organisation depending on the nature and the location of the project.
Screening	The competent authority makes a decision on whether EIA is required. This may happen when the competent authority receives notification of the intention to make a development consent application, or the developer may make an application for a screening opinion. The screening decision must be recorded and made public.
Scoping	The EU directive provides that developers may request a scoping opinion from the competent authority. The scoping opinion will identify the matters to be covered in the environmental information. It may also cover other aspects of the EIA process.
Environmental studies	The developer carries out studies to collect and prepare the environmental information required.
Submission of Environmental Information to Competent Authority	The developer submits the environmental information to the competent authority together with the application for development consent. The environmental information is presented usually in the form of an environmental impact statement.
Review of adequacy of the environmental information	The client may be required to provide further information if the submitted information is deemed to be inadequate.
	(cont'd overleaf)

Key stages	Notes
Consultation with statutory environmental authorities, other interested parties and the public	The environmental information must be made available to authorities with environmental responsibilities and to other interested organisations and the general public for review. They must be given an opportunity to comment on the project and its environmental effects before a decision is made on development consent.
Consideration of the environmental information by the competent authority before making development consent decision	The environmental information and the results of consultations must be considered by the competent authority in reaching its decision on the application for development consent.
Announcement of decision	The decision must be made available to the public including the reasons for it and a description of the measures that will be required to mitigate adverse environmental effects.
Post-decision monitoring if project is granted consent	There may be a requirement to monitor the effects of the project once it is implemented.

Establishing whether EIA is required

Generally, it will fall to local planning authorities in the first instance to consider whether a proposed development requires EIA. For this purpose they will first need to consider whether the development is decried in Schedule 1 or Schedule 2 to the regulations. Development of a type listed in Schedule 1 always requires EIA. Development listed in Schedule 2 requires EIA if it is likely to have significant effects on the environment by virtue of factors such as its size, nature or location.

Development which comprises a change or extension requires EIA only if the change or extension is likely to have significant environmental effects.

Like the Town and Country Planning Act, the Regulations do not bind developments by Crown bodies.

Planning applications

Where EIA is required for a planning application made in outline, the requirements of the regulations must be fully met at the outline stage since reserved matters cannot be subject to EIA. When any planning application is made in outline, the local planning authority will need to satisfy itself that it has sufficient information available on the environmental effects of the proposal to determine whether or not planning permission should be granted in principle.

Where the authority's opinion is that EIA is required, but not submitted with the planning application, it must notify the applicant within three weeks of the date of receipt of the application, giving full reasons for its view clearly and precisely.

An applicant who still wishes to continue with the application must reply within three weeks of the date of such notification. The reply should indicate the applicant's intention either to provide an environmental statement or to ask the secretary of state for a screening direction. If the applicant does not reply within three weeks, the application will be deemed to have been refused.

Preparation and content of an environmental statement

It is the applicant's responsibility to prepare the ES. There is no statutory provision as to the form of an ES. However, it must contain the information specified in Part II, and such of the relevant information in Part I of Schedule 4 to the regulations as is reasonably required to assess the effects of the project and which the developer can reasonably be required to compile. (See Appendix 3.)

The list of aspects of the environment which might be significant affected by a project is set out in paragraph 3 of Part I of Schedule 4 (see Appendix 4), and includes human beings, flora, fauna, soil, water, air, climate, landscape, material, assets, including architectural and archaeological heritage and the interaction between any of the foregoing.

Procedures for establishing whether or not EIA is required ('screening')

The determination of whether EIA is required for a particular development proposal can take place at a number of different stages:

■   The developer may decide that EIA will be required and submit a statement.

■   The developer may, before submitting any planning application, request a screening opinion from the local planning authority. If the developer disputes the need for EIA (or a screening opinion is not adopted within the required period), the developers may apply to the secretary of state for a screening direction. Similar procedures apply to permitted development.

■   The local planning authority may determine that EIA is required following receipt of a planning application. If the developer disputes the need for EIA, the applicant may apply to the secretary of state for a screening direction.

■   The secretary of state may determine that EIA is required for an application that has been called in for his determination or is before him on appeal.

■   The secretary of state may direct that EIA is required at any stage prior to the granting of consent for particular development.

Provision to seek a formal opinion from the local planning authority on the scope of an ES ('scoping')

Before making a planning application, a developer may ask the local planning authority for their formal opinion on the information to be supplied in the ES (a 'scoping opinion'). This provision allows the developer to be clear about what the local planning authority considers the main effects of the development are likely to be and, therefore, the topics on which the ES should focus. The developer must include the same information as would be required to accompany a request for a screening opinion. The local planning authority must adopt a scoping opinion within five weeks of receiving a request.

Provision of information by the consultation bodies

Under the Environmental Information Regulations, public bodies must make environmental information available to any person who requests it. Once a developer has given the local planning authority notice in writing that he intends to submit an ES, the authority must inform the consultation bodies. The consultation bodies are:

■ The bodies who would be statutory consultees under article 10 of the General Permitted Development Order for any planning application for the proposed development.

■ Any principal council for the area in which the land is situated (other than the local planning authority)

■ English Nature

■ Scottish Natural Heritage

■ Countryside Council for Wales

■ Northern Ireland Environmental Agency

■ Joint Nature Conservation Committee

■ Council for Nature Conservation and the Countryside

■ The Environment Agency

Selection criteria for screening Schedule 2 development

This is a reproduction of Schedule 3 of the regulations (paragraphs 20 and 33):

1. **Characteristics of development**

   The characteristics of development must be considered having regard, in particular, to:

   i.   the size of the development

   ii.   the cumulation with other development

   iii. the use of natural resources

   iv.   the production of waste

   v.   pollution and nuisances

   vi.   the risk of accidents, having regard in particular to substances or technologies used.

2. **Location of development**

   The environmental sensitivity of geographical areas likely to be affected by development must be considered, having regard, in particular, to:

   i.   the existing land use

   ii.   the relative abundance, quality and regenerative capacity of natural resources in the area

   iii. the absorption capacity of the natural environment, paying particular attention to the following areas:

   a.   Wetlands

   b.   coastal zones

   c.   mountain and forest areas

   d.   nature reserves and parks

e. areas classified or protected under Member States' legislation; areas designated by Member States pursuant to Council Directive 79/409/EEC on the conservation of natural habitats and of wild fauna and flora

f. areas in which the environmental quality standards laid down in Community legislation have already been exceeded.

g. densely populated areas

h. landscape of historical, cultural or archaeological significance.

3. **Characteristics of the potential impact**

The potential significant effects of development must be considered in relation to criteria set out under paragraphs 1 and 2 above, and having regard in particular to:

i. the extent of the impact (geographical area and size of the affected population)

ii. the transfrontier nature of the impact

iii. the magnitude and complexity of the impact

iv. the probability of the impact

v. the duration, frequency and reversibility of the impact.

Information to be included in an environmental statement

This is a reproduction of Schedule 4 of the regulations (paragraphs 81–85 and 91).

Part I

1. Description of the development, including in particular:

i. a description of the physical characteristics of the whole development and the land-use requirements during the construction and operational phases

ii. a description of the main characteristics of the production processes, for instance, nature and quantity of the materials used

iii. an estimate, by type and quantity, of expected residues and emissions (water, air and soil pollution, noise, vibration, light, heat, radiation, etc.) resulting from the operation of the proposed development.

2. An outline of the main alternatives studied by the applicant or appellant and an indication of the main reasons for his choice, taking into account the environmental effects.

3. A description of the aspects of the environment likely to be significantly affected by the development, including, in particular, population, fauna, flora, soil, water, air, climatic factors, material assets, including the architectural and archaeological heritage, landscape and the inter-relationship between the above factors.

4. A description of the likely significant effects of the development on the environment, which should cover the direct effects and any indirect, secondary, cumulative, short, medium and long term, permanent and temporary, positive and negative effects of the development, resulting from:

    i.   the existence of the development

    ii.   the use of natural resources

    iii. the emission of pollutants, the creation of nuisances and the elimination of waste

and the description by the applicant of the forecasting methods used to assess the effects on the environment.

5.   A description of the measures envisaged to prevent, reduce and where possible offset any significant adverse effects on the environment.

6.   A non-technical summary of the information provided under paragraphs 1–5 of this part.

7.   An indication of any difficulties (technical difficulties or lack of know-how) encountered by the applicant in compiling the required information.

Part II

1.   A description of the development comprising information on the site, design and size of the development.

2.   A description of the measures envisaged in order to avoid, reduce and, if possible, remedy significant adverse effects.

3.   The data required to identify and assess the main effects which the development is likely to have on the environment.

4.   An outline of the main alternatives studied by the applicant or appellant and an indication of the main reasons for his choice, taking into account the environmental effects.

5.   A non-technical summary of the information provided under paragraphs 1–4 of this part.

The characteristics of a good environmental impact assessment

■   A clear structure with a logical sequence, for example, describing, existing baseline conditions, predicted impacts (nature, extent and magnitude), scope for mitigation, agreed mitigation measures, significance of unavoidable/residual impacts for each environmental topic.

■   A table of contents at the beginning of the document.

■   A clear description of the development consent procedure and how EIA fits within it.

■   Reads as a single document with appropriate cross-referencing.

■   Is concise, comprehensive and objective.

■   Written in an impartial manner without bias.

■   Includes a full description of the development proposals.

■   Makes effective use of diagrams, illustrations, photographs and other graphics to support the text.

■   Uses consistent terminology with a glossary.

■   References all information sources used.

■ Has a clear explanation of complex issues.

■ Contains a good description of the methods used for the studies of each environmental topic.

■ Covers each environmental topic in a way which is proportionate to its importance.

■ Provides evidence of good consultations.

■ Includes a clear discussion of alternatives.

■ Makes a commitment to mitigation (with a programme) and to monitoring.

■ Has a non-technical summary which does not contain technical jargon.

Sample environmental statement

1.0 INTRODUCTION

1.1 General Project Description
1.2 EIA Development
1.3 Planning Context
1.4 Scope and Content of the ES
1.5 ES Availability and Comments

2.0 EIA METHODOLOGY

2.1 Objectives
2.2 Scoping Study
2.3 Consultations
2.4 Defining the Baseline
2.5 Sensitive Receptions
2.6 Impact Prediction
2.7 Evaluation of Significance
2.8 Mitigation
2.9 Residual Impact
2.10 Assumptions and Limitations

3.0 DEVELOPMENT BACKGROUND AND ALTERNATIVES

3.1 Introduction
3.2 Site Considerations and Constraints
3.3 No Development alternative
3.4 Objectives of the Proposed Redevelopment
3.5 Design Alternatives

4.0 THE SITE DESCRIPTION AND DESIGN STATEMENT

4.1 Introduction
4.2 Site Location and Setting
4.3 Site Description
4.4 The Design Statement

5.0 PLANT DISMANTLING, DEMOLTION, REMEDIATION AND CONSTRUCTION

5.1 Introduction
5.2 Schedule Overview
5.3 Plant Dismantling and Asbestos Removal
5.4 Demolition
5.5 Remediation
5.6 Construction
5.7 Construction Traffic
5.8 Environmental Management Plan and Code of Construction

6.0 ENVIRONMENTAL MANAGEMENT PLAN (EMP) AND POTENTIAL IMPACTS OF THE CONSTRUCTION WORKS

6.1 Introduction
6.2 Scope of the EMP
6.3 Summary and Conclusions

7.0 PLANNING AND POLICY CONTEXT

7.1 Introduction
7.2 Planning Policy Guidance
7.3 Strategic Guidance
7.4 Strategic Planning in (location)
7.5 Planning Brief
7.6 The Adopted local UDP
7.7 UDP Proposed Alterations
7.8 Affordable and Social Housing
7.9 Summary and Conclusions to Planning and Policy Context

8.0 SUSTAINABILITY – ENVIRONMENTAL

8.1 Introduction
8.2 National Guidance and Local Policy
8.3 Approach to Assessment
8.4 Sustainability Topics
8.5 Results
8.6 Summary and Conclusions

9.0 SOCIO-ECONOMIC IMPACTS

9.1 Introduction
9.2 Approach to Assessment
9.3 Baseline Statistics
9.4 Impact of the Development
9.5 Summary and Conclusions

10.0 BUILT HERITAGE, TOWNSCAPE AND VISUAL IMPACTS

10.1 Introduction
10.2 Approach to Assessment
10.3 Approach to Presentation of the Visual Assessment
10.4 Baseline Condition – The Heritage and Existing Townscape
10.5 Townscape Studies, Policies and Guidelines
10.6 Townscape and Visual Impact Assessment
10.7 Impact Assessment – The Immediate Locality
10.8 Impact Assessment – The Panorama, The Moving Eye and the Interaction of Major Built Forms
10.9 Impact Assessment – Specific Viewpoints Reviewed
10.10 Summary and Conclusions

11.0 ARCHAEOLOGY

11.1 Introduction
11.2 Approach to Assessment
11.3 Policy Considerations and Legislative
11.4 Initial Assessment
11.5 Archaeological Potential of the Site
11.6 Environmental Potential of the Site
11.7 Archaeological Resources in the Surrounding Area
11.8 Summary of Archaeological Potential
11.9 Impact of the Development

Part 1 Appendices

APPENDIX 12 Application of project scheduling software

At its lowest level, the software sold for scheduling as 'project management software' can be no more than a drawing tool or, at its highest, a complex arrangement of customisable relational databases and graphical front end. In order to be capable of producing a schedule that can perform as a time model, the software must have an adequately functional relational database at its core. The reason for this is that the software has to be capable of computing the consequences of change, while a drawing tool, which simply illustrates the decisions made by the drafter, cannot do this.

Many schedulers are trained by the software manufacturers to operate the software they use. This is extremely important and useful training. However, it is not training in time management and should not be thought of as a substitute for it. By analogy, many of us have experience of securing a good grounding on how Microsoft Word works, but even with its spelling and grammar checking the software will not guarantee that what is written is useful, technically accurate or even intelligible.

No matter how high the quality of the software, it cannot produce a high-quality output of its own accord. Even the best project planning software will not secure the competent management of time.

While every company considering software products will from time to time wish to take into consideration matters peculiar to themselves, or matters peculiar to the project on which they wish to work, there are certain considerations that should transcend subjective preferences and there are certain software attributes which are desirable for the purposes of competent time management. Because software changes by the day and 'new and improved' products (which unfortunately is a term often confused with 'more bells and whistles') are brought to the market, those attributes that are desirable only for the purposes of time management, irrespective of whether they are currently available in any particular product, are listed below.

It is unhelpful if different parties to a particular project use different software because different products work in different ways and even if given the same data will produce different calculations from different algorithms. Accordingly, all parties to a project should use the same software and a departure should not be permitted.

While getting to grips with unfamiliar software may be tedious, competent schedulers will generally know their way around any new product within an hour and will rarely take more than a day or two to become sufficiently capable of using it competently. Unfamiliarity with scheduling software products should thus not be a serious consideration in product selection.

Primary software considerations

Projects and subprojects

Software that can only cope with a single project at a time is unlikely to be sufficiently flexible for complex projects. For example, apart from the possibility of identifying as subprojects separate sections that are subject to sectional completion or separate key dates, for ease of application in practice, it may also be useful to identify separate operational zones as subprojects.

Activities

For each activity there should be: (a) a unique activity-identifying alphanumeric code and (b) a unique description. Software that permits duplication of activity identifiers (IDs) or activity descriptions without warning is likely to produce schedules that lack clarity and are thus incompatible with good practice. The software should not facilitate that duplication at all, or, if it does, have a clear permanent warning on the schedule as to the deficiency.

The software should be capable of distinguishing between the following activity and event types:

- duration-identified activities

- resource-calculated activities

- hammocks

- start milestones or flags

- finish milestones or flags

- employer-owned contingency/risk allocation

- contractor-owned contingency/risk allocation.

An activity-related field capable of taking free-text and numbers as comments or notes is often a useful facility. The software should be capable of identifying activity durations in different formats. Although for most purposes in construction, activity durations in days may be sufficient, for the purposes of limited possessions durations in hours and minutes and in outline schedules, durations in weeks and months are necessary. The software should make it clear to what unit of time it carries out its calculations, i.e. days, hours, minutes or seconds. The best software calculates to the minute.

The software ought to be capable of identifying which activities are logically determined to be of a shorter duration than the applied logic and whether they are to be 'stretched' or 'not continuous' as a result of the logic, or the logic changed.

Logical relationships

The software should permit a logical flow of work and prohibit the indication of relationships which are impossible to perform. Any software that fails to do that is likely to produce schedules that are incompatible with good practice and should not facilitate it at all, or declare a clear permanent warning on the schedule as to the deficiency.

The software should be capable of identifying all variations of logical links either individually or in combination. Software that limits the user to finish-to-start logic or few logical connections to any one activity is unlikely to be useful.

The software should identify any inconsistency between logic and the activity durations to which the logic is applied. Logic should be capable of being illustrated as 'driving' or 'non-driving' to any chosen point within the model. Logic should distinguish between:

- engineering logic (the construction sequence with no resource constraints)

- resource logic (the construction sequence carried out with the available resources)

- preferential logic (the construction sequence with imposed constraints to modify the purely 'engineered' and/or 'resourced' construction sequence)

- logic linking zones and or subprojects.

The software should be capable of identifying fixed lead and lag and the working calendar the lead or lag is to adopt. Lead and lags should be listed as logic attributes.

Constraints

Manually applied constraints are likely to be useful on most projects. Those which are acceptable, when correctly applied are:

- start no earlier than a given date

- start no later than a given date

- start as late as possible (also known as zero-free float).

The software should be capable of clearly identifying when a manual constraint has been applied to an activity. Some software facilitates the use of constraints that will manipulate criticality and inhibit the ability of the software to accurately model time. These are seldom acceptable in a schedule used to manage time and which, if available in a software product, should not be permitted to be used without a clear permanent warning on the schedule as to their effect. Possible constraints are:

- any combination of constraints that will fix the earliest and latest dates for any activity or milestone

- a mandatory start date

- a mandatory finish date

- zero-total float.

Critical path

The software should be capable of identifying:

- the longest path to completion

- the longest path to intermediate key dates or sectional completion dates

- logic and activities that are critical from those that are not critical to one or more completion dates

- total float on each path

- free float on each activity on each path.

The software should be capable of facilitating the tracing of a critical path or paths through the driving logic of each activity on the critical path to a particular completion date or key date from time to time.

Calendars

The software should be capable of facilitating the use of a number of different working calendars for activities, resources and lags, each capable of identifying:

- working week start day

- working weeks and weekends
- working days
- working hours
- holidays
- standard calendars and exceptions.

Resources

The software should be capable of facilitating the use of a number of different resources and determining whether or not the duration of the activity to which they are allocated is to be calculated by reference to the specified resources in terms of:

- name of resource
- unit working period
- allocated working calendar
- number of units of work per calendar period
- cost per unit
- availability.

Work breakdown structure and activity coding

The software should be capable of identifying a work breakdown structure. While a structure of eight levels should be the ideal, a structure of less than five levels is unlikely to be practical on a complex project. The facility for a broad variety of bespoke database fields and can be displayed are usually an essential requirement of complex schedules.

Organisation

The software should be capable of organising the layout in any combination of fields and attributes, sorting on activity and logic and attributes and values in any field.

Filtering

The software should be capable of filtering the content of any layout to select the value of any field, or attribute, either alone or in combination with other fields, or attributes on the basis of the value or attribute:

- equal to
- containing
- not equal to
- not containing

and, where the field contains a value, there should also be the facility in relation to those values for such for a selection falling: between or not between.

Layout

The minimum available layouts should comprise the following:

- bar chart without logic

- bar chart with logic

- network diagram (ADM or PDM)

- resource profile

- cost profile.

The software should have the facility for creating and saving a variety of different combinations of fields and attributes organised and filtered, as layouts for reporting purposes. The time scale to which the layout is restricted to view should be identifiable to any duration and density during the period between six months prior to inception of the earliest project and 12 years after planned completion of the latest project.

Every layout should be printable as both hard copy and a portable document format (PDF).

As-built data

The software should be capable of identifying the factual data for each activity and resource as:

- actual duration

- start date

- finish date

- percentage complete

- remaining duration

- calculated cost

- actual cost

- certified value

- resources expended.

Updating

The software must be capable of identifying a data date by a straight line through the activity bars at that date. The software must be capable of comparing progress against the currently agreed baseline, such that any delays and/or changes in activity sequencing are clearly demonstrated.

The software must be capable of recalculating the critical path or paths and the predicted early and late start and finish dates of all activities and resources against the data date with the effect that:

- all activities indicated to have started or finished are indicated to have started or finished earlier than the data date

- no activity is identified to have started or finished later than the data date

■ activities that are in progress at the data date are indicated to be due to finish on a date after the data date proportionate to their degree of progress in relation to their planned duration at the data date.

Inputting and editing data

The software should be capable of holding input data and edits in memory so that they are subject to undo and only saved on a positive instruction to do so.

Archiving

Files should be capable of being saved in compressed data format for archival purposes.

Training and support

The availability of effective, product-related training is extremely useful even for experienced schedulers. Even with the simplest of software, it is always helpful to understand what the software supplier identifies as the way it should be used. Because of the sophistication of modern software and the inability, or unwillingness, of the manufacturers to subject products to rigorous testing before release, committed and easily available software support and continuous updated software is more important today than it has ever been.

Secondary software considerations

Those matters which do not add to the quality of the calculated output but to the manner of use and which, depending on circumstances, may be of some importance are listed below.

Enterprise-wide software

Enterprise-wide software that can directly link a project or projects across the internet such that large, complex projects are able of being scheduled and effectively monitored across the world. Enterprise-wide software that can relate all projects together with which the company is concerned. It is a useful attribute for enhancing company management.

Communications

Whether the schedule can be accessed by other parties via the internet, in whole or in part, for viewing only, or for editing, with secure access rights can be of importance in managing the schedule.

Appearance

Software capable of being customised according to the company requirements for house style by using different fonts, line thickness or type and colours for each available field or value in the database and the background is useful.

A drawing facility that can be used to highlight aspects of a report is also often useful.

Comparison of schedules

For the purpose of identifying the effect of differences between schedules in the process of review, revision, updating and impacting causative events, it is useful to have the facility for comparing two or more schedules on a line-by-line basis. In practice this usually means the facility for identifying one or more target schedules that can be viewed simultaneously with the current schedule.

Organisation

A facility for organising the layout in order of logical predecessors and successors is useful.

Transparency with other software

The facility for importing from and exporting to other scheduling software may be available but if it is, it should be capable of listing the differences that result from such import or export.

Integration with time- and cost-keeping systems can facilitate automatic updating from time, plant and material records which, in relation to a fully resourced schedule can produce an automated update facility.

The facility for attaching hyperlinks to activity IDs should be available to facilitate the linking of such documents as photographs and movies, flow charts, procedures and method statements and progress records.

Risk analysis

The facility for stepping through a potential shift in timing of activities to ascertain the consequent shift in the critical path is useful. Monte Carlo analysis will give a profile of likelihood of success against given criteria which, if accurately predicted against data that remains unchanged, will predict likely outcome.

Archiving

A backup capable of being set to default periods or to be executed manually is a useful facility.

APPENDIX 13 Change management

Change in a construction project is any incident, event, decision or anything else that affects:

- the scope, objectives, requirements or brief of the project

- the value (including project cost and whole-life cost) of the project

- the time milestones (including design, construction, occupation)

- risk allocation and mitigation

- working of the project team (internally or externally)

- any project process at any project phase.

Changes during the design development process

The procedure outlined is used to control the development of the project design from the design brief to preparation of tender documents. It will include:

- addressing issues in the design brief

- variations from the design brief, including deign team variations and client variations

- developing details consistent with the design brief

- approving key design development stages, namely scheme design approval and detailed design approval.

The procedure is based on the design development control sheet. The approved design will comprise the design brief and the full set of approved design development control sheets. The procedure comprises the following stages:

- The appropriate member of the design team addressing each design issue in the development of the brief, co-ordinated by the design team leader.

- Proposals developed are discussed with the appropriate members of the project's core group through submission of detailed reports/meetings co-ordinated by the project manager. Reports should not repeat the design brief, but expand it, address an issue and prepare a change.

- The design team leader co-ordinates preparation of a design development control sheet, giving:
  - design brief section and page references
  - a statement of the issue
  - a statement of the options
  - the cost plan item, reference and current cost
  - the effect of the recommendation on the cost plan and the schedule
  - a statement as to whether the recommendation requires transfer of client contingency (i.e. a client variation to the brief) and if so the amount to be transferred.

- The design team report section of the control sheet is signed by:
  - the design team member responsible for recommendations
  - the quantity surveyor (for cost effect)

○　　the design team leader (for co-ordination).

■　The design team leader sends the design development control sheet to the project manager who obtains the client's approval signature and returns it to the design team leader.

■　The quantity surveyor incorporates the effect of the approved recommendation into the cost plan.

■　The project manager incorporates the effect of the approval recommendation into the master schedule.

Design development control sheet

Client name:		
Project name:		
Sheet no.		

Design team report

Design brief section:

Issue: Pages:

Options considered: 1.
                           2.
                           3.

Recommendation:

Cost plan item:

Ref:

Current cost:

Effect of recommendation on costs/schedules: Increase/decrease

Application for transfer of client contingency: Yes/No   Amount:

Architect/services engineer/structural engineer: Date:

Quantity surveyor: Date:

Design team leader: Date:

Client approval

Design development/Client contingency transfer approved (delete as applicable)

Position   Signature   Date

Example of change management process

- Identification of requirement for change.

- Evaluation of change.

- Consideration of implications and impact including risks.

- Preparation of change order.

- Reviewing of change order: client's decision stage.

- Implement change.

- Feedback including causes of change.

Change order request form

Project no.	Date:	No.	
Client: **Project:**			**Distribution:**
Subject – definition of change:			WHAT
Identified by:			WHO
Reasons for change:			WHY
	Discretionary		Non-discretionary	
Cost implication:			
Time implication:			
Recommended action:		Project manager	.........Date
Client decision required by:		Date:	
Forwarded to client:		Date:	
Client's decision:		Date	
Projected schedule and cost plan (budget) amended on		Project manager	

Part 1 Appendices

Change order register

Project:			Client:		Job Id:	File reference:		
Request Id:	Date	Initiated by:	Description of change	Client decision required by:	Client decision obtained by:	Client decision	Client decision Id no.:

APPENDIX 14 Procedure for the selection and appointment of consultants

Stages	Key steps
Strategy	■ Decide works procurement strategy
	■ Prepare project brief
	■ Prepare consultant's brief
	■ Decide terms of engagement including the choice of single/multiple appointment and phased appointment
Pre-selection	■ Prepare preliminary list
	■ Decide criteria of selection
Selection	■ Invite to tender
	■ Evaluate tender
	■ Assess tender
Appointment	■ Finalise terms of engagement
	■ Finalise management, monitoring and review process

Guidance for selection process

1. Determine what duties are to be assumed by the consultants and prepare a schedule of responsibilities. If applicable, consider what level of in-house expertise is available.

2. Check to see if the client has any in-house procedures or standard conditions of engagement for the appointment of consultants and what scope there is for deviating from them.

3. Decide on the qualities most needed for the project, and the method of appointment. Agree them with the client.

4. Establish criteria for evaluating consultants with weighting values (e.g. 5 vital, 0 unimportant) for each criterion.

5. Assemble a list of candidates from references and recommendations. Check any in-house approved and updated lists of consultants.

6. Prepare a shortlist by gathering information about possible candidates. Check which firms or individuals are prepared in principle to submit a proposal.

7. Assess candidates against general criteria and invite proposals from a select number (no more than six or less than three per discipline). Invitation documents should be prepared in accordance with the checklist given below. Competitive fee bids, if required, should conform to relevant codes of practice.

8. Arrange for conditions of engagement to be drawn up. The conditions, the form of which will vary with the work required and the type of client, should refer to a schedule of responsibilities for the stages for which the consultant is appointed and include a clause dictating compliance with the project handbook. The conditions of engagement should be based as closely as possible on industry standards (e.g. as set by CIC, RIBA, ACE, NEC and RICS).

Consistency of style and structure between conditions for different members of the team will improve each member's understanding of their own and others' responsibilities. Each set should include this aspect. Fee calculation and payment terms should be clearly defined at the outset, together with the treatment of expenses, i.e. included or not in the agreed fee.

9. Determine the criteria for assessing the consultants' proposals. Agree them with the client.

10. Appraise the proposals and select the candidates most appropriate for the project. Proposals should be analysed against the agreed criteria using weighting analysis.

11. Arrange final interview with selected candidates (minimum of two) for final selection/negotiation as necessary.

12. Submit a report and recommendation to the client.

13. Client appoints selected consultant.

14. Unsuccessful candidates are notified that an appointment has been made.

Checklist

1. The Consultant's brief must include:

   - project objectives
   - requirements of other participants
   - services to be provided
   - project schedule including the key dates
   - requirement of reports including key dates.

2. Invitation documents must include:

   - a schedule of responsibilities
   - the form of interview panel
   - draft conditions of engagement (an indication of the type to be used)
   - design skills or expertise required
   - personnel who will work on the project, their roles, time-scales, commitment, output
   - warranties required, for whose benefit and in what terms.

Invitations should ask candidates to include information on the level of current professional indemnity insurance cover for the duration of the project. Details of policy, date of expiry and extent of cover for subcontracted services must be provided.

Example of consultancy services at different project stages

The following is a brief list of consultancy services typically provided at different stages of a project. However, this neither is a comprehensive list nor outlines the preferred sequence as that may vary from project to project. For a detailed scope of services, documents such as the CIC Scope of Services can be consulted.

Inception/feasibility

- Identification of client requirements, objectives and a commitment to sustainability including preparation of the project brief.

- Feasibility studies including evaluation of options, environmental impact assessment, site assessment, planning guidance and commercial assessment.

Strategy/pre-construction

- Design development including preparation of outline design and scheme design.

- Development of cost estimates, tender preparation and evaluation and preparing project schedule.

- Preparation of construction specifications and schedules.

Construction/commissioning

- Preparation and issuing working drawings and variations.

- Project/construction management.

- Inspection, monitoring and valuation of construction.

- Certification of payment.

- Advising on dispute resolution.

- Confirmation of completion.

- Assisting in project handover.

Completion/handover

- Ensuring defects correction.

- Settlement of project including final accounts.

- Confirmation of operation and maintenance procedures.

- Post-project appraisal and feedback.

Part 1 Appendices

APPENDIX 15 Characteristics of different procurement options

	Characteristic	Traditional	Design and build	Management contracting	Construction management
1.	Diversity of responsibility	Moderate	Limited	Large	Large
2.	Size of market from which costs can be tested	Moderate	Limited	Moderate	Large
3.	Timing of predicted cost certainty	Moderate	Early	Late	Late
4.	Need for early precise definition of client requirements	Yes	Yes	No	No
5.	Availability of independent assistance in development of design brief	Yes	No	Yes	Yes
6.	Speed of mobilisation	Slow	Fast	Fast	Fast
7.	Flexibility in implementing changes	Reasonable	Limited	Reasonable	Good
8.	Availability of recognised standard documentation	Yes	Yes	Yes	Limited
9.	Ability to develop proposals progressively with limited and progressive commitment	Reasonable	Limited	Reasonable	Good
10.	Cost-monitoring provision	Good	Poor	Reasonable	Good
11.	Construction expertise input to design	Moderate	Good	Good	Good
12.	Management of design production programme	Poor	Good	Good	Good
13.	Client influence on trade contractors	Limited	None	Good	Good
14.	Provision for controlling quality of construction materials and workmanship	Moderate	Moderate	Moderate	Good
15.	Opportunity for contractor to exploit cash flow	Yes	Yes	Yes	No
16.	Financial incentive for contractor to manage effectively	Strong	Strong	Weak	Minimal
17.	Propensity for confrontation	High	Moderate	Moderate	Minimal

The significance of the features listed in the table above may be outlined as follows:

1.   Having widely dispersed responsibilities for different activities may provide the project manager with greater control, e.g. in the selection of preferred consultants. It may, however, make it difficult to pinpoint responsibility.

2.   It is acceptable practice to limit the numbers of tenders invited from contractors based on value and design development criteria. Where tendering involves significant design development the cost will discourage contractors unless invitations are restricted. Such restrictions may not produce the most competitive price available unless careful pre-tender assessments are made.

3.   Although the establishment of a certain financial outcome at an early stage in the development programme will minimise client risks it could well be at a

Part 1 Appendices

price. This is because of the risks which the tenderers will have to assume. A balance has to be achieved that depends on all the circumstances.

4. The client's requirements document associated with design and build procurement is a definitive statement. It must be produced early and it becomes the basis for all subsequent activities. Other procurement options enable progressive development of the client's brief, which may be helpful where there is uncertainty or greater complexity.

5. Independent assistance with the development of a design brief, which is integral to procurement options, may be advantageous where there is uncertainty or greater complexity, similar to item 4.

6. Mobilisation of construction using traditional procurement is relatively slow because much of the design development must be completed before appointment of the contractor, whereas all other methods enable progressive design and construction.

7. Little flexibility to accommodate variations exists within the design and build method. The other methods make reasonable provision for flexibility through the issuing of variations or additional works contracts.

8. Standard documentation, with which the industry is familiar, allows agreements to be entered readily. Although it enables incorporation of particular requirements the drafting of unique documents often involves much negotiation and expense.

9. Where there are significant uncertainties or where limited finance is available, the opportunity to develop and appraise proposals may be advantageous. There may even be an opportunity to carry out construction on a progressive basis, step by step.

10. All procurement methods should seek to provide facilities for client cost monitoring, although the possible detail will vary.

11. Contractors' input to design could produce more cost-effective solutions provided contractors' interests are accommodated correctly. By using design and build the contractor clearly has a vested interest in providing such input.

12. The schedule of preparation of production information is often critical to, and should be determined by, the construction schedule.

13. Procurement methods have different abilities to select preferred trade or works contractors that actually execute the works; no influence in selection is possible using design and build, and only limited influence is possible using the traditional method.

14. Design and build procurement makes no provision for the monitoring of construction quality – any monitoring required by the client must be independently commissioned. In other forms of procurement members of the design team – or the management contractor or the construction manager – may have monitoring responsibilities. But in all cases, except the last, only limited control of quality is available.

15. Design and build standard forms usually make no provision for the provision of a working schedule. Accordingly, it is essential to effective project control that the necessary amendments are made to the standard form to facilitate the control of time in accordance with the CIOB's *Guide to Good Practice in the Man-*

agement of Time in Complex Projects and to enable the project manager to properly execute his responsibilities in that regard.

16. Since construction works involve substantial financial transactions there is considerable financial benefit to the main contractor in achieving payments as promptly as possible while delaying payments due as long as possible. This may have a significant detrimental effect on the attitudes and performance of the specialist subcontractors, and hence on the quality of their workmanship, thus exacerbating the limited quality control characteristics of procurement methods. Where payments to specialist contractors are under direct control of the client/construction manager, this can be turned to advantage.

17. Management procurement methods provide for remuneration of the management contractor or construction manager on the basis of a fee, not necessarily related to performance. Equitable performance measurement is often difficult. In design and build procurement there is a strong incentive for good management; in traditional procurement there is also an incentive for the contractor.

18. Construction quality, speed and cost can all be improved through good teamwork. Procurement methods which recognise the varying responsibilities of those managing construction operations and which preclude exploitation of any party are most likely to avoid confrontation.

Selection of the procurement method

From the foregoing it can be seen that the most important characteristics of each method will best suit particular types of project. For example, design and build procurement would be an obvious choice where a client has limited interest in involvement with the design or construction process; and when there are clearly defined, straightforward requirements, including a need for early determination of cost.

It is necessary to consider all the characteristics of the project and to compare them with the characteristics of the various procurement methods available. The most important characteristics should be identified initially, after which secondary and peripheral issues should be considered, and the details determined for any necessary adaptation of the basic procurement methods available. For example, although design and build procurement may appear well suited to the project characteristics, it will probably be appropriate for the client to appoint an architect or planning consultant to progress the project through planning approval. The documentation produced would then be incorporated in the client's requirements on which design and build tenders would be sought.

Care must be taken in adapting any particular procurement method to compensate for perceived shortcomings, to avoid compromising the basic principles and essential characteristics. Thus, for example, although the engagement of design assistance for preparation of the client's requirements will inevitably dilute the single-point responsibility attribute of design and build procurement, the effects of this dilution should be mitigated by careful definition of responsibilities and terms of engagement. Similar care must be exercised when procuring specialist components or services incorporating design elements within a traditional project procurement method.

Selection of a procurement method is thus an essential element in the development of the policies to be adopted for implementation of all projects. In view of the fundamental differences in philosophy between the four basic procurement methods, the method should be determined at the earliest possible stage so that timely decisions can be made on the engagement of appropriate project resources. The development process can be optimised only by giving consideration at the earliest stage to the issues on which the appropriate procurement method should be determined.

Selecting a procurement route

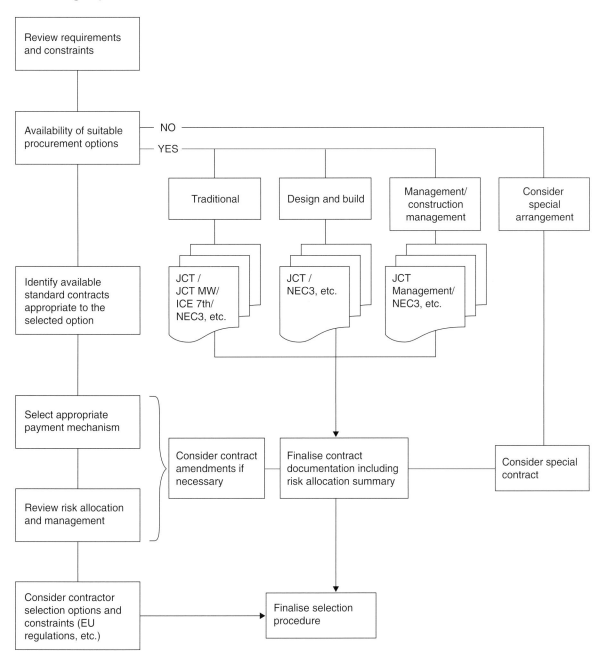

APPENDIX 16 Dispute resolution methods[1]

The construction industry is a fertile breeding ground for disputes. They cannot be avoided entirely and it would be foolish to suggest that they could. Among other things, there may be design faults, there can be defective work or materials, the cost of variations may cause dismay, money can be wrongfully withheld, loss and expense for delay and prolongation, or extensions of time to defend against liquidated damages for late completion may be claimed.

On the other hand, the high cost of energy-sapping defended litigation can often be avoided by sensibly planning your dispute resolution procedures before contract and as well as by the proactive management of the process of resolution once a dispute has arisen.

Mediation, conciliation, expert determination, adjudication, arbitration, and, of course litigation, are all possibilities to be considered. Two of these: mediation and conciliation are often referred to as alternative dispute resolution (ADR). That in itself does not mean much without recognising to what it is an alternative. The essential difference between orthodox dispute resolution and ADR is that in ADR the parties make their own settlement agreement, which is only binding so long as they want it to be, and in orthodox dispute resolution the decision is made for them by a third-party and it is final and binding on them. There is a grey area in all this and that is in expert determination and adjudication in which the decision can be final and binding, or it can be final and binding unless disputed in another forum, or it could be non-binding depending on how (and under what law) it is structured.

Apart from reference to the courts by litigation (which in every common-law country is a unilateral act, open to anyone who thinks they have had a right infringed) all the other methods of dispute resolution require an agreement. Naturally, it is easier to agree a method of resolving a dispute before it has arisen rather than after. However, irrespective of whether there is an agreement in place, it is always open to either party to suggest an alternative means of dispute resolution that will save both parties time, cost, and frustration, and to enter into an agreement for that, at any time.

Non-binding

In non-binding processes the dispute resolver helps the parties to agree their differences. These are entirely private processes, conducted without prejudice to the rights of either party and there is nothing stopping either party from shifting its ground during the process. Indeed, if it is to be successful, it is essential that they do. If they do not succeed in reaching a settlement there is nothing to prevent either party from dealing with the same dispute through another forum at a later date and nothing that has been discussed in the ADR process may be used in evidence can usually be used elsewhere. The dispute resolver will agree a procedure with both parties; he will read the parties' respective position statements and any documents provided in support. He will consult with the parties privately, and with both together. Although essentially a non-binding process, it is always open to both parties to agree that the final settlement should be binding. The parties agree to share the costs of the dispute resolver and to pay their own. This is an excellent method of dealing with disputes, because it encourages the parties to talk to each

---

[1] Extracted from Pickavance, Keith, 'Dispute resolution without tears', Times of The Islands, (Spring, 2005). http://www.timespub.tc/2005/04/transforming-waste-into-wonder/ (accessed 31 July 2009).

other; if successful it helps to preserve working relationships and, even if unsuccessful, it helps the parties to focus on the real matters in which they are in dispute. In many contracts, ADR is required at some stage and in England, court-ordered ADR forms a part of the civil procedure rules of the courts.

Mediation

Without express permission, the mediator will never disclose what has been said to him by either party to the other. A mediator does not have to have a detailed understanding of the facts or the law of the matters in dispute, but it often helps. He will not advise the parties of their rights nor generally will he advise the parties of the strength of their case, but he will help each to see the weaknesses of their own and the strengths of their opponent's position. In doing so he will draw them closer with a view to executing an agreement to settle their differences. In general, mediation can be completed in two to three days. In very large cases with many issues it might take a week or more but that is unusual.

Conciliation

Conciliation as a similar process to mediation but the conciliator takes a more active role in the settlement of the dispute than does the mediator. A conciliator necessarily has to have to a detailed understanding of the facts and law of the matters in dispute. The conciliator will express an opinion on the relative merits of the parties' respective cases. He will try to persuade them of his views and, in doing so, will attempt to guide the parties into an agreement compatible with the parties' rights under the contract. Conciliation can be expected to be a little shorter than mediation simply because the conciliator is able to focus the parties' attention on the issues and drive the process in a way that is unavailable to a mediator. In general, conciliation can be completed in one or two days. As with mediation, in very large cases with many issues it might take a week or more but that is unusual.

Non-binding or final and binding

Unlike ADR, in which the parties make their own decision, the essence of these decision-making processes is that a third party is introduced to make the decision for them. Because the process is consensual, it is always a private process. However, depending on the rules of engagement agreed between the parties, the information that becomes available may not be privileged and the decision made may not be binding on the parties, leaving them free to revisit the dispute in another forum. The parties are free to agree who should pay the dispute resolver's costs and how the party's costs should be dealt with, although it is usual for each side to pay their own costs.

Expert determination

Expert determination is quite different from any other method of dispute resolution. In this forum the expert is appointed for his knowledge and understanding of the particular issues in dispute in the field in which he is an acknowledged expert. The expert will agree a procedure with both parties; he will read the parties' respective position statements and any documents provided in support. There is usually no provision for the parties to change their position or amend their case during the process. He will consult with the parties privately, and may consult with both together, but he is under no obligation to do so unless it is made a term of his appointment. He is given the role of investigator. He is required to find the facts and law in relation to the issues in dispute, to make his own enquiries, tests and calculations and to form his own opinion and decide on the merits of the parties' position. Depending on the issues, expert determination can involve much research and a hearing and can take anything from a week to several months.

Adjudication

In England and Wales and in several Commonwealth countries, adjudication has recently been given statutory authority. Under the law of those countries that adopt this process it is generally the rule that either party to a specified type of construction contract has the right at any time to submit any dispute or difference to the adjudication of a third party. However, even where the statutory right is limited to particular types of contract there is nothing stopping the parties from agreeing by contract to follow the same process in regard to contracts which are outside the law.

Adjudicators are often appointed for their knowledge and experience of the type of matters in dispute, although it is not essential. While the idea of adjudication is that there should be a decision, in the event that the parties do not like the result there is nothing to prevent them from running the case again in another forum; the rule of *res judicata* does not apply to adjudication. The adjudicator will agree a procedure with both parties; he will read the parties' respective referrals and any documents provided in support. He may also require a hearing and will often conduct conference calls with the parties.

The adjudicator's decision is binding until either party decides to refer the same dispute to arbitration or litigation, in which case the decision is binding until an award or judgment is handed down. When the legislation was first enacted in England in 2000, the adjudicator was empowered to make his own enquiries of the facts and law. It was thought that he might act pretty much like an architect or engineer under a construction contract and that few parties would take the adjudicator's decision as final and binding, so it was not initially thought necessary for the adjudicator to act within the rules of natural justice.

Five years on and several hundred enforcement cases later, it became clear that parties who have been unhappy with the outcome have sought to overturn the decision on the basis of the adjudicator's misconduct rather than have the case rerun in arbitration or litigation. As a result the courts have imposed the obligation on adjudicator's to act within the rules of natural justice. They must hear both sides. The parties must have an equal opportunity to make their own case and to respond to the case against them, although they may not alter or amend their submissions. This is a tall order in the limited time available to make the decision. They must be impartial but they do not have to be independent. They may only enquire into the facts and the law of the cases that are put to their decision. They may not go outside the parameters of the parties' submissions to make good any deficiencies.

Unless the referring party agrees to extend the period for the decision by up to 14 days, or both parties agree to extend the period the decision beyond that the dispute resolution process must be conducted and the decision given within 28 days of referral. The adjudicator has no power to order discovery or to take evidence on oath unless the parties give it to him by agreement and if either party request it, he must give reasons for his decision. It all seems to work very satisfactorily.

Part 1 Appendices

Final and binding

In the following tribunals the facts once found cannot be reopened by any court: the matters are *res judicata*. Appeal on a point of law is always available from a domestic arbitral tribunal to the court and from a lower court to a higher court. However, statute has tended to limit the right of appeal from an arbitrator's award in other than a point of law of public importance in order to give the parties a greater sense of finality.

Arbitration

An arbitration agreement is written into all standard forms of building and civil engineering contract. It is a private process and nobody is permitted to know of the matters in dispute or the decision unless the parties agree otherwise. The arbitrator's decision is final and binding and can be enforced in many countries by virtue of the New York convention. Arbitrators, like judges, must be independent and impartial. They must scrupulously follow the law of the contract and the rules of natural justice to provide a speedy and efficient decision on all the issues submitted to jurisdiction. The arbitrator may not go outside that limitation to decide things that were not part of the reference.

Subject to the arbitration agreement, the parties may adopt specific procedural rules which dictate the powers of the arbitrator or the procedure to be followed. If this is not done, the powers of the arbitrator are set out by statutory instrument. In domestic disputes it is normal for the reference to be to a single arbitrator, but in international disputes it is more common for each party to appoint their own arbitrator and for the arbitrators to appoint a chairman or umpire, forming a three-man tribunal.

Arbitration can be very time consuming and expensive or it can be quick and cheap depending on the parties and the case-management skills of the arbitrator. There is usually nothing to stop a party from amending its case subject to paying the costs of the other side. Generally, the arbitrator has the powers of a high court judge in regard to the taking of evidence on oath, subpoenas for evidence, discovery and so on. He can order a party to pay the costs of interlocutory matters and can determine who should pay his fees and whether the losing party should pay the winning party's costs, in whole or in part, with or without interest and on what basis. The arbitrator must give reasons for his decision if either party requests it.

Litigation

Litigation is the dispute resolution process run by the civil courts of the state. It is free to every individual who has a grievance to resolve. Judges tend not to be technical people although in some courts they are specifically selected for their technical ability (e.g. the English Technology and Construction Court). On the other hand, judges often have the power to appoint technical assessors or experts to assist them and will almost always do so if the parties request it.

Notwithstanding that the court and the judge are provided by the state, litigation is often a very expensive process. This is often simply because of the complicated rules of procedure, which a reluctant but wily litigant can often exploit to put off the hearing of the case for years, including amending its case from time to time. There are also restrictions on who can appear in the courts on behalf of a litigant. In large cases the costs can run to many thousands of pounds per day during a hearing, which may take many months or even years before the dispute reaches that stage.

Litigation is a public process (justice must be seen to be done) and the public are encouraged to sit in on the proceedings to hear of the matters in dispute. Judges must give reasons for their decisions and important decisions are published and recorded in law reports.

APPENDIX 17 Regular reports to the client

Notes for guidance on contents

Executive summary

The purpose of the executive summary is to give the client a snapshot of the project on a particular date which can be absorbed in a few minutes. It should contain short precise statements on the following:

■ Significant events that have been achieved.

■ Significant events that have not been achieved and action being taken.

■ Significant events in the near future, particularly where they require specific action.

■ Progress against the master, design and construction schedules.

■ Financial status of the project.

Contractual arrangements (including legal agreements)

Each project requires the client to enter into a number of legal agreements with parties such as local authorities, funding institutions, purchasers, tenants, consultants and contractors. The report should be subdivided to identify each particular agreement and to provide details of requirements and progress made against the original project master schedule. The following are indications of possible legal agreements that may be required on a project:

■ joint development agreement

■ land purchase agreement

■ funding agreement

■ purchase agreement

■ tenant/lease agreement

■ consultants' appointments

■ Town and Country Planning Acts: sections in force at the time, e.g.

    ○ planning gain

    ○ highways agreement

    ○ planning notices

    ○ land adoption agreement

    ○ public utilities diversion contracts.

Client's brief and requirements

This provides a 'status' report on how the client's brief and requirements are progressing. The report should identify any requirements which need clarification or amplification and also those which are still to be defined by the client.

Client change requests

Client-orientated changes should be listed under status (being considered, in progress, completed), cost and schedule implications. The objective is to make the client fully aware of the impact and progress of any change.

Planning building regulations and fire officer consents

This section will be subdivided into the various consents required on a specific project. Each section should highlight progress made, problems possible solutions and action required or in progress. The following are examples of possible consents:

■ planning – outline

■   planning – detailed, including conditions

■   Building Regulations

■   means of escape

■   English Heritage/Historic Buildings

■   fire officer

■   public health

■   environmental health

■   party-wall awards.

Public utilities   Each separate utility should be dealt with in terms of commitment, progress, completion and any agreements, way-leaves as appropriate.

Design reports – summaries   The design team and consultants should prepare reports on progress problems and solutions which will form the appendices and must include marked-up design schedules and 'issue of information' schedules. The design report, however, should be distilled into an 'impact-making' synopsis and agreed as a fair representation by each member concerned.

Health and safety   Report on the preparation of the CDM health and safety plan and the health and safety file.

Project master schedule   Updated schedules should form an appendix to the report, specifying progress made. A short commentary on any noteworthy aspects should be made under this section.

Tendering report   This is a status report on events leading up to the acceptance of tenders. It should show clearly how the various stages are progressing against the action plan.

Construction report summary   This report is prepared in a similar way to that outlined for design reports (above).

Construction schedule   The updated schedule should form an appendix to the report, highlighting progress made and showing where delays are occurring or are anticipated. A short commentary on any important items should be given in the report under this section.

Financial report   A fully detailed financial report should form one of the appendices. It should provide a condensed overview (say two to three pages) giving the financial status and cash flow of the project. This report will embrace the information provided by the quantity surveyor and also call for the project manager to provide an overall financial view, highlighting any specific matters of interest to the client.

Appendices   These will include full reports and schedule updates as outlined in the previous sections. Other reports, possibly of a specialist nature, may also be included. Should the report be presented at a formal meeting then the minutes of previous meetings should be included in the appendices.

APPENDIX 18 Practical completion checklist

Project no:`

Project Management Ltd'

Authorised to approve   Signature

_____   _____

_____   _____

Have/has the following been completed?

1. Contract works. ☐

2. Commissioning of engineering services. ☐

3. Outstanding works schedule issued. ☐

4. Outstanding works completed. ☐

5. Operating and maintenance manuals, 'as built' drawings and C&T records issued ☐

6. Maintenance contracts put into place. ☐

7. Building Regulations consent signed off. ☐

8. Occupation certificate issued. ☐

9. Public health consent signed off. ☐

10. Health and safety consent signed off and health and safety file available. ☐

11. Planning consent complied with in full, including reserved matters. ☐

12. Equipment test certificates issued (lifts, cleaning cradle, others). ☐

13. Insurers' certificates issued (lifts, cleaning cradle, sprinklers, others). ☐

14. Means of escape signed off. ☐

15. Fire-fighting systems and appliances signed off. ☐

16. Fire alarm system signed off and fire certificate issued. ☐

17. Public utilities way-leaves and lease agreements signed off. ☐

18. Public utilities supplies inspected and signed off. ☐

19. Licences to store controlled chemicals. ☐

20. Licences to dispose of controlled chemicals. ☐

21. Licences to store gases. ☐

22. Licence to use artesian well. ☐

23. Adoption of highways, estate roads, and walkways by local authorities. ☐

24. Consent to erect and maintain flag-poles. ☐

25. Consent to erect illuminated signs. ☐

26. Cleaning to required standard. ☐

27. Removal of unwanted materials and debris. ☐

28. Tools and spares. ☐

29. Client/user insurances established. ☐

Completed	/
not applicable	X

APPENDIX 19 Facilities management

Facilities management (FM) started out as property management primarily concerned with the management of premises. As commercial reality and competitiveness demanded greater efficiency, attention focused on the need to manage not just the buildings but the entire resources used by organisations in the generation of their wealth, hence FM. It is not a new concept but one that has progressed from use by a handful of companies to become the fastest-growing property and resource management sector in construction.

FM seeks to create a framework that embraces the traditional estate management functions of property maintenance, lighting and heating with increasingly analytical reviews of space occupation/planning, asset registers, health and safety registers, and activity flow throughout the premises. Hence, the term 'facilities' is used to include all the buildings, furnishings, equipment and environment available to the workforce while pursuing the company's business goals.

The success of FM has been greatly enhanced by the development of reliable and powerful computer technology together with the boom in personal computers that has made serious data handling affordable to all. The use of databases to control the occupational activities of buildings is both reactive and proactive with the latter gaining in importance. The reactive use allows data on the performance of the workplace to be collected and stored, this in turn is available for historical analysis that can be used proactively to identify recurring trends and anticipate operational problems, so eliminating waste.

Every FM application in industry and commerce is in effect a one-off system; it must address the priorities of the company but is actually assembled from a series of independent modules that operate from a universal FM platform. The emerging industry-standard platform is based on the computer-aided design technology used extensively in the design of buildings; this has been developed into powerful computer-aided facilities management (CAFM) systems. CAFM systems are increasingly likely to serve as an indispensable source of reference for the project manager and project team in drawing up the project brief for buildings of similar function. The pairing of FM and project management in this way should enable the procurement of increasingly efficient property.

The CIOB has become aware of the vast array of bespoke, tailor-made FM contracts which are prevalent in the FM industry. Many of these contracts are often based on models from other industries, and suffer from lack of focus on FM contractual issues.

CIOB, in partnership with Cameron McKenna, published the first standard form of FM contract in 1999. The third edition of this document had been published in 2008.

APPENDIX 20 Value for money project framework

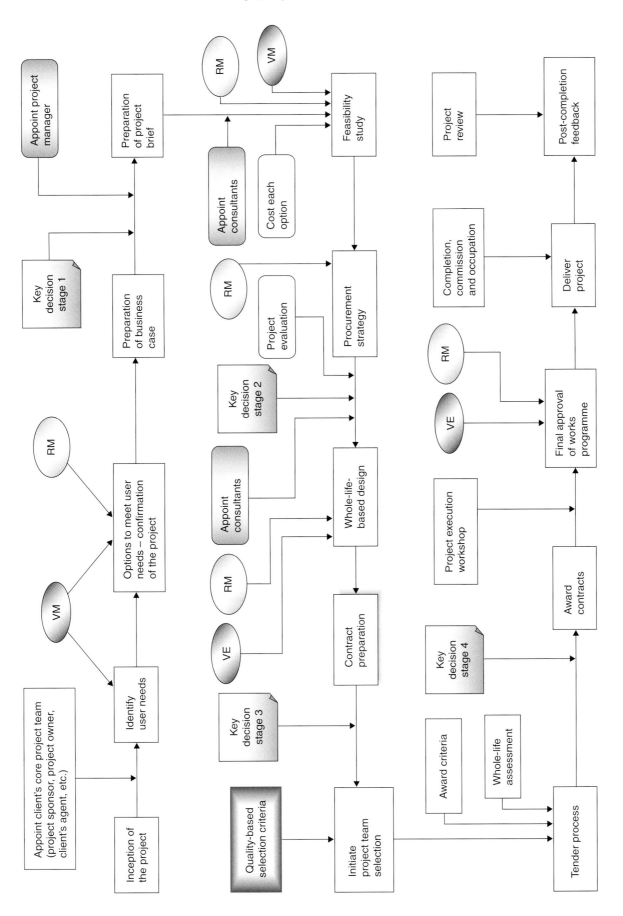

APPENDIX 21 Leadership in project management

What is leadership?

Leadership, as a management attribute have been subjected to a significant amount of attention. Defining simplistically, 'it is the process in which an individual influences other group members towards the attainment of group or organisational goals' (Shackleton[1]). Inevitably, there are a wide range of theories and schools of thoughts encompassing this subject (a number of reference documents are listed in the bibliography). The latest discussions tend to focus on transactional and transformational natures of leadership.

Leadership and project manager

The very definition of construction project management implies that within a defined timescale, the project is expected to achieve an agreed set of targets utilising specific resources. This requires not only a very efficient project manager but also an effective project leader who can lead the project team spontaneously, mainly focusing towards the project and motivating the project team members to achieve the targets within the agreed project framework. The key aspects or traits that a successful project manager would require to excel in are motivation, performance appraisal, resource allocation and management, and planning and communication.

What are the traits of effective leaders?

There are a range of theories present outlining leadership styles and traits. Broadly, successful leadership traits are characterised in six styles as detailed in the table below.

Leadership styles[2]

Style	Result
Coercive	Leader demands immediate compliance
Authoritative	Leader mobilises people towards a vision
Affiliative	Leader creates emotional bonds and harmony
Democratic	Leaders use participation to create consensus
Pacesetting	Leader expects excellence and self direction from the group
Coaching	Leader develops people for the future

The general suggestion is that the leaders need to understand how the styles relate to their individual competencies and situational requirements so as to identify the most suitable approach.

There are some differences of views on effectiveness of training of leadership skills (are leaders born or made?). However, it is advisable to stress the need for flexibility of the leader – to learn to lead differently depending on the situational and contextual needs; hence the leaders should learn many styles and learn to diagnose the needs of the context and situation.

---

[1] Shackleton, V. (1995) *Business Leadership*. London: Routledge
[2] Goleman, D. (2000) Leadership that gets results. *Harvard Business Review*, March–April.

Are there any quick wins?

Although different styles and tactics would suit different contexts and situations, the adaptation of the following should enhance effective project management:

■ acknowledging positive contribution

■ ensuring open communication

■ 'touching base' with the team members on a regular basis

■ sharing praise.

Part 1 Appendices

APPENDIX 22 Framework agreements

Framework agreements can be described as agreements to provide both goods and services on predefined and specified terms and conditions with a selected number of suppliers (e.g. consultants, designers and constructors). While entering into a framework agreement does not by implication necessitate a binding requirement on either side to procure or provide the goods or services, the framework agreement will usually specify the terms that will apply if and when they are provided. There are a wide variety of framework procurement contracts including forms published by JCT, NEC and OGC.

The current EU Directives (2004/18/EC1 – Article 32) and the UK Regulations (enacted 31 January 2006 Regulation 19) also expressly provide for this form of procurement. It is to be noted that a framework agreement is different from a framework contract.

What are the advertising requirements?

The OJEU contact notice, as a minimum, must:

- make it clear that a framework agreement is being awarded

- include the identities of all the contracting authorities entitled to call-off under the terms of the framework agreement

- state the length of the framework agreement; usually it will be a maximum of four years

- include the estimated total value of the goods, works or services for which call-offs are to be placed and, so far as is possible, the value and frequency of the call-offs to be awarded.

How is the framework agreement awarded?

- Open, restricted or, in certain circumstances, negotiated or competitive dialogue processes can be used.

- Award can be made to one or multiple providers (at least three).

- Mandatory standstill period applies – only for the agreement – not for the future call-offs.

What is the process for a call-off?

- When awarding individual contracts under framework agreements (called call-offs), authorities do not have to go through the full procedural steps in the EU Directives again.

- The weighting criteria can be varied subject to certain conditions, i.e. the criteria used for mini-competition can be different from those used for the framework award.

- Call-offs may not need mini-competitions if the terms laid down in the framework agreements are sufficiently precise. However, how this is to be done is not set out in the regulations.

- Not all suppliers need to be included in the mini-competitions, particularly if the framework is split in different 'lots'.

The process charts for framework agreements and call-offs are set out below:

Framework agreements

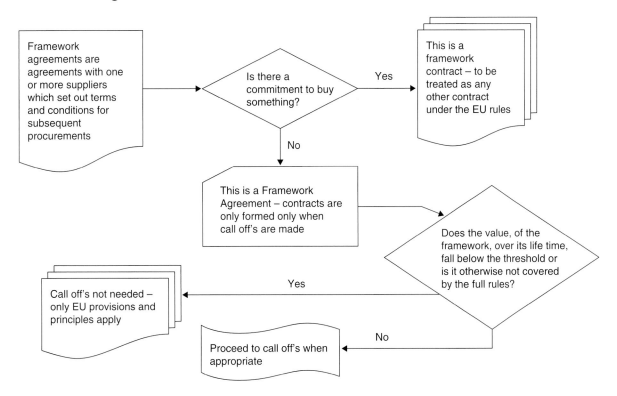

Call-off stage

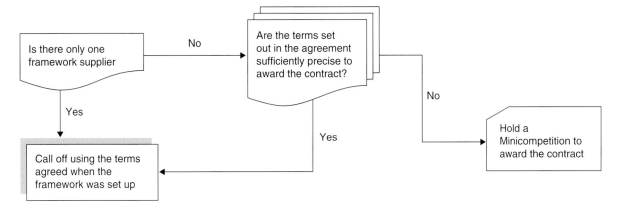

APPENDIX 23 Guidance on e-procurement

e-Procurement in the public sector

A recent Office of Government Commerce (OGC) publication indicated that of an estimated five million purchase orders being placed by central government each year, almost 30% are now being processed using e-procurement systems. The publication further noted that departments and agencies placed significant focus on using electronic auctions (in which suppliers compete online by improving their price offer to buyers during a short 'live' online session). This led OGC to conclude that e-auctions have proved relatively easy to implement and can provide impressive 'quick win' savings averaging over 13%, with IT hardware savings achieving as much as 23%.

In a survey of private sector business, the Department for Trade and Industry found that UK companies considered that one of the key roles for technology was to: 'drive supply chain efficiencies and to unlock value for the buyer'. The Office for National Statistics' own research drew similar conclusions. Of the 7000 companies polled, it found that: 'those using eCommerce for purchasing have a significant productivity advantage over those who don't'. The message from public and private sector alike is clear: 'e' stands for efficient, effective and empowered and its use can generate significant savings.

e-Procurement and Europe

Some of the key e-procurement participant countries in Europe include Austria, Belgium, Denmark, France, Germany, Ireland, Italy, Norway, Portugal and Sweden. There are many successful e-procurement solutions within Europe. However, at time of going to print there is no single government organisation in Europe that has implemented a comprehensive suite of electronic tools and systems to support all public procurement activity. Comparatively, the UK government has made admirable progress in this area. Considerable investment and commitment to e-procurement continues across Europe.

EU Directives

The EU encourages the use of e-procurement. The new EU Consolidated Directives and EU Invoicing Directives make clear provision for the use of electronic tools and techniques within public sector purchasing across Europe. The EU acknowledges that automating processes and enabling opportunities to be advertised and tendered online fully supports the aims and objectives of cross-border trading, non-discrimination and fair and open competition. They are also encouraged by the transparency and the ease of auditing online transactions. Innovative tools such as electronic reverse auctions are provided for within the new Directives, and the EU has even gone a step further by introducing a new process, Dynamic Purchasing Systems, an online-only process whereby suppliers can compete for contracts.

e-Procurement best practice

The 'quick wins' approach

To modernise procurement processes, certain 'e-tools' will be required. One problem often encountered at the start of an e-procurement programme is deciding which tools to implement and in what order. 'Quick wins' can establish the credibility of the e-procurement programme, and help to generate funding for the rest of the programme. Examples of these can include use of procurement cards [similar to the Government Procurement Card (GPC)], e-auctions, e-sourcing and similar initiatives.

Implementing e-procurement

A wide variety of e-procurement tools have been developed over recent years to help organisations source, contract and purchase more efficiently and effectively. Broadly, e-procurement tools relate to two aspects of procurement: sourcing activity and transactional purchasing.

Sourcing activity (e-sourcing)

The e-sourcing tools can help buyers establish optimum contracts with suppliers and manage them effectively. The tools include supplier databases and electronic tendering tools, evaluation, collaboration and negotiation tools. Also included are e-auction tools and those tools which support contract management activity.

Transactional purchasing (e-purchasing)

The e-purchasing tools can help procurement professionals and end users achieve more efficient processes and more accurate order details. The two aims of maximising control and process efficiency are the function of e-purchasing tools such as purchase-to-pay systems, purchasing cards and electronic invoicing solutions. Although the tools fall broadly within these two categories, some tools can be implemented in isolation.

e-Auction tools are now a mature technology that can generally be implemented more quickly than other e-sourcing tools. As e-auctions are currently proving a clear 'quick win' in cash-releasing terms, their earliest implementation is strongly recommended. The diagram below shows where e-sourcing and e-purchasing (called purchase to pay) tools fit in the procurement lifecycle as classified by the Chartered Institute of Purchasing and Supply (CIPS).

CIPS e-Procurement lifecycle

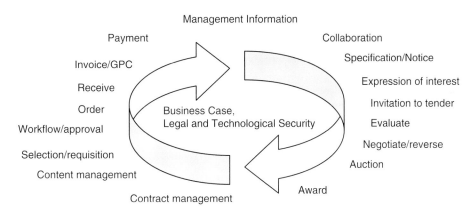

Purchasing cards (P-cards)

Purchasing cards (P-cards) are similar in principle to charge cards used by consumers (e.g. suppliers are paid within five days; the buyer is billed monthly in a consolidated invoice), but with extra features which make them more suitable for business-to-business purchasing. These can include: controls such as restricting card use to particular commodity areas, individual transaction values and monthly expenditure limits. The purchasing information provided to the buying organisation by an issuing bank on each monthly statement depends on the degree of detail automatically generated by each supplier. This can range from the supplier name, date and transaction value, to line item detail against each item ordered, free text entry for the input of account codes and VAT values.

Supplier participation in P-card programmes

Many suppliers already accept consumer credit and debit card payments and no extra equipment is required to accept P-cards. The costs to suppliers in accepting credit, debit and P-card payments are a small transaction charge (normally ranging from 1% to 4%) and the cost of implementing the card-processing equipment. This cost increases with the higher level capabilities.

Benefits of P-cards

■ Process savings.

■ Prompt payment discounts.

■ Guaranteeing prompt payment.

■ Increased compliance with contracts.

e-Auctions

In an electronic reverse auction (e-auction) potential suppliers compete online and in 'real time', providing prices for the goods/services under auction. Prices start at one level and gradually, throughout the course of the e-auction, reduce as suppliers offer improved terms in order to gain the contract. e-Auctions can be based on price alone or can be weighted to account for other criteria such as quality, delivery or service levels.

e-Auction benefits

Government e-auction activity shows an average saving of 13.4% over previous contract value. Further benefits include improved preparation and planning for the tendering process, opportunity for suppliers to submit revised bids for a contract (as opposed to the formal tendering process), and increased market knowledge for buyers and suppliers. Suppliers particularly benefit from increased awareness of competitor pricing.

Implementing e-auctions

e-Auctions do not replace tendering: they are a part of it and provide cost-effective, fast and transparent conclusions to a full tendering process. e-Auctions may be based on securing the lowest price, or on most economically advantageous bid (price, payment terms, supply schedules). Only those suppliers who have successfully pre-qualified (i.e. they have satisfied all tendering criteria such as quality processes, financial stability and environmental policies) should be invited to participate. The complexity of an organisation's procurement will affect the e-auction strategy. Some basic considerations for all requirements, whether complex or simple, are:

■ **Starting price**: what will be the starting price criteria? For example, an indicative price submitted by suppliers in an earlier stage of the tendering process?

■ **Bid decrements**: what will be the minimum level by which a supplier can reduce their bid below the current lowest? For a £100,000 contract a bid decrement of £2000–5000 would be reasonable.

■ **Duration:** what will be the duration of the event?

■ **Extensions:** what extensions will be granted? For example, if any bids are received within the last five minutes of the e-auction an extension of five minutes might be granted for other bidders to respond.

■ **Weightings:** more complex e-auctions will allow suppliers to revise their bids in respect of criteria including, but not restricted to, price.

Further information can be available from various sources including SIMAP and OGC. Also advice must be sought from the individual service providers.

Suppliers and e-auctions

Suppliers are generally co-operative about participating in e-auctions. Buyers should maintain excellent communications with suppliers, being open and providing all relevant operational and technical information (this is a legal requirement). Buyers should also provide supplier inductions to e-auctions and a test e-auction prior to the live event if necessary to ensure supplier familiarity with both the process and the technology.

Each supplier may adopt a different strategy for participation: some bidding lower prices early, others holding back. Suppliers may wish to see where they stand in the bidding process and the value of other bids, but not the names of other bidders.

Use of e-auctions in the construction industry

The UK public sector is implementing e-auctions as a valuable tool for improving the purchasing process. There is now more experience in both the public and private sectors to demonstrate that e-auctions have been found to improve professionalism, speed up the process and, in many cases, reduce the purchase price for goods and services. Within the UK public sector, e-auctions are being implemented in accordance with best practice, supported by a professional code of conduct. e-Auctions form only one stage of a full-quality tender process. e-Auctions themselves can be complex with weighted options to take account of factors other than price throughout the process. This approach ensures that contracts continue to be awarded on a value for money basis and not on price alone in line with government policy.

The construction industry has been progressing with adoption of e-commerce in the same way as other sectors. However, there have been some strong objections to electronic reverse auctions (e-auctions) from some sections of the industry. OGC has received representations from trade associations and other bodies. Sections of the industry have seen e-auctions as a return to lowest price purchasing, threatening already low margins. The industry also perceives e-auctions as challenging the principles of the Achieving Excellence in Construction initiative, such as an integrated supply chain approach to construction procurement based on optimum whole-life value.

Part 1 Appendices

APPENDIX 24 Guidance on good practice contract management framework

The Office of Government Commerce (OGC) and the National Audit Office have jointly developed a best practice guidance on contract management framework (particularly relevant to the pre-construction and construction stages) based on the lessons learned on various projects. The salient aspects of this good practice advice are summarised below (full details can be found on the OGC website).

Structure/resources	**Planning/ governance**	■ Planned and systematic transition from procurement to contract management ■ Clear ownership for contract management ■ Defined and clear contract management plan with a whole life approach to performance
	People	■ Continuity ideally through involvement of the contract manager during procurement ■ Ensuring appropriate knowledge and skill availability ■ Clear expectations and requirements from all those involved in contract management ■ Encouraging knowledge and good practice sharing
	Administration	■ Keeping documentation accessible, secure and up to date ■ Ensuring adequate resources are available for information and documentation management ■ Ensuring triggering mechanisms for key contract elements (e.g. notice periods) ■ Availability of regular and 'as and when' reporting of contract management information
Delivery	**Relationships**	■ Clear, visible and defined roles and responsibilities for all involved ■ All parties understand their individual roles and mutual expectations and obligations ■ Incentives are available to encourage and nourish positive relationships
	Continuity and communications	■ Appropriate handover process from procurement stage ■ Regular and 'as and when' communication routes established ■ Stakeholders and kept up to date with progress (with newsletters or similar) so as to encapsulate and manage expectations ■ Ensuring communication plan is agreed in advance and in place including measures for problem resolution and relationship management

	Managing performance	■ Structured and mutually understood delivery process ■ Performance management framework is in place ■ Service level agreements are in place ■ Performance reporting includes self-assessment as well as independent checking ■ Clear processes are in place to expedite operational problem resolution ■ Regular and routine performance feedback is provided to suppliers ■ Clear contact points established for all parties ■ Changes are captured and processed formally
	Payment and incentives	■ Payment mechanisms are documented and are clear and understood by all parties ■ Payment processes are well defined and efficient with appropriate checks and balances being in place ■ Costs are monitored against budget and allocated appropriately ■ Incentive structures relate clearly to the desired outcomes ■ When open book or similar mechanisms are used, the process is managed professionally and fairly
	Risk	■ Contractual including supplier risk management is in place with clear responsibilities and processes ■ Identification is carried out of who is best placed to manage risk and allocation is made accordingly ■ Escalation and reporting routes are in place including contingency plans ■ Risk register is maintained and actioned diligently
Development	**Change management**	■ Regular review is undertaken to identify and action change requirements ■ Change management process is clear and structured – with differentiating minor and major changes ■ All parties concerned understand the change management process ■ Appropriate dispute resolution processes are in place
	Supply chain development	■ Supply chain performance improvement activities are undertaken where appropriate ■ Shared risk reduction initiatives are undertaken ■ Shared management activities may be initiated to obtain performance improvements

Strategy	**Supply chain relationship**	■ A structured supplier relationship management programme is in place with clear expectations and anticipated outcomes ■ Focus is on capturing innovation where necessary or appropriate ■ Knowledge is shared and captured ■ Stakeholder interfaces and relationships are planned and managed
	Market management	■ Processes are in place to evaluate and review options for delivering services in house or out-sourced ■ Understanding is gained of the market including the suppliers ■ Capacity and capability of potential suppliers is analysed and understood ■ Processes exist to continually monitor available opportunities ■ Feedback is provided for strategy development of the 'new' procurement process

APPENDIX 25 Communication plan

Most projects, particularly those of complex nature and with multiple stakeholders will have communication strategies that take into consideration the client, design team, delivery team and the internal and external stakeholders. This is discussed and is agreed at the pre-construction stage to ensure that all parties are satisfied with the process, nature and frequency of communications. Particular attention should be paid to identify the communication focal points for each party who in turn would spearhead the spread of the information.

A typical project communications outline may encompass a wide variety of organisations, an example of which is shown in the diagram.

Relationship Diagram

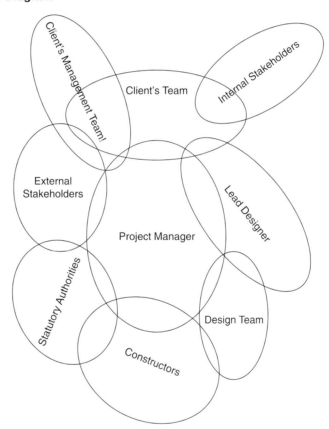

The development of the communication plan should focus on facilitating the process of keeping the key stakeholders informed of the project's progress and to promote 'buy-in' by making the project's development visible at all times. There are many ways in which the relationship with, and perceptions of the internal and external stakeholders can be managed. Some of the key tools that can be used are:

- communications meetings: engaging a wider range of stakeholders

- relationship workshop: engaging both internal and external stakeholders

- newsletters: highlighting the progress of the project and forthcoming actions.

In addition to these, particularly at the construction stage, facilities such as web-cams can be used, if appropriate, to provide either real-time or time-delayed progress that can be transmitted either to a selective audience or to the public in general through the internet.

Initiatives such as Considerate Constructors Scheme also encourage various campaigns and techniques to enhance project communications to a wider range of stakeholders.

APPENDIX 26 Good practice project management[1]

Focus on long-term cost	■ Is the evaluation based on whole-life value for money, taking into account of capital, maintenance and service costs? ■ Does the evaluation process balance cost, quality and time? ■ Are there adequate checks and balances to ensure fair application of the evaluation process?
Engage, understand and utilise the supply chain	■ Do the suppliers understand the project delivery approach and agree that it is achievable? ■ Have the supplier assumptions made against their proposals been checked and understood? ■ Are the supply-side risks assessed? ■ Are there processes in place to ensure that all parties have a clear understanding of their roles and responsibilities? ■ Is there a shared understanding of desired outcomes, key terms and deadlines?
Adopt an integrated approach	■ Has the market been tested for responsiveness to the project requirements? ■ Are the procurements routes being used encouraging integration of the project team? ■ Are the suppliers being engaged early to help determine and validate the project outcomes and outputs? Is there incentive for value management? ■ Has a shared risk register been established?

Establish clear links between project and strategic priorities	■ How does the project compare and align with the business and/or organisational delivery and operational activities? ■ Are the lessons learned from previous projects being applied? ■ Is there a clear project plan that covers the full period or project delivery, based on realistic time scales, showing critical dependencies and how delays can be handled? ■ Has an analysis been undertaken of the effects of any slippage in time, cost, scope or quality? In the event of a problem and or conflict does the project team realise that at least one must be sacrificed?
Identify and agree measures of success for each stage of project lifecycle	■ Has the key success factor for each stage been identified and agreed? ■ Has a process been put in place to measure the success factors? ■ Have the success factors been agreed with stakeholders and suppliers relevant to each stage? ■ Does the project team have a clear view of the interdependencies between projects, the benefits and the criteria against which success will be judged?

[1] Information distilled from guidance available from the Office of Government Commerce – Good Practice Contract Management Framework

Establish ownership, support and leadership	■ Are all the proposed commitments and targets first checked for delivery implications? ■ Are decisions taken early, decisively and adhered to, in order to facilitate successful delivery? ■ Have all the key actions and risks been allocated to the most suitable party? ■ Has coherence been established in the delivery team?
Manage stakeholders	■ Are there clear governance and communication arrangements to suitable alignment of the objectives of all the stakeholders? ■ Have the right stakeholders been identified for all the stages? ■ Has the rationale for the stakeholders (e.g. the why, the what, the who, the where, the when and the how) been identified? ■ Has a common understanding and agreement of stakeholder requirements been secured? ■ Does the business case take account of the views of all stakeholders including users? Is there a clear understanding of the process to manage stakeholders (e.g. ensures buy-in, overcomes barriers and resistance to change and allocate risk to the party best able to manage it)? ■ Has sufficient account been taken of the subsisting organisational cultures? ■ While ensuring that there is clear accountability, has a process been established to resolve any conflicting priorities?
Utilise project management and risk-management techniques	■ Is there a skilled and experienced project team with clearly defined roles and responsibilities? ■ Are the major risks identified, weighted and treated? ■ Have sufficient resources (financial or otherwise) been allocated to the project, including an allowance for risk? ■ Are there adequate approaches for estimating, monitoring and controlling cost? ■ Are there effective systems for monitoring and measuring the realisation of the benefits in the business case? ■ Are the governance processes robust enough to ensure that 'bad news' is not filtered out of the reporting system? ■ Are all the team members accountable and committed to help to ensure successful and timely delivery?
Ensure realistic targets with key decision points	■ Has sufficient time been built in to allow for statutory approvals and lead-ins? ■ Have the delivery timescales been kept short so that change during development is avoided? ■ Are there adequate review points built in so that the project scope/time/cost/specification can be changed if circumstances change? ■ Is there a business continuity plan in the event of the project delivering late or failing to deliver at all?

APPENDIX 27 Compliance with Site Waste Management Plan Regulations 2008

Introduction

The Site Waste Management Plan Regulations (derived from Clean Neighbourhoods and Environment Act 2005) has been enacted from April 2008 which enforced the legal requirement for preparing a site waste management plan (SWMP) prior to commencement on a site for all construction and demolition projects in England, with an estimated project cost of more than £300,000.

It is estimated that the UK construction industry uses 400 million tonnes of resources every year with 100 million tonnes ending up as waste. This equates to approximately 1.65 tonnes of waste for every person living in the UK. Site waste is harmful to the environment and to the economy of the industry. Research by CIRIA estimates that almost 13% of all materials delivered to a site end up in a skip without ever being used. With the rising cost of sending waste to landfill, the incentive to manage waste more efficiently is not only ethical but also essentially commercial value addition.

What is a SWMP?

A SWMP provides a framework for managing the disposal of waste throughout the life of a construction project. The SWMP will identify estimated quantities of waste that a site will produce at the pre-construction stage using information based on the design and facilitate the identification of the optimum decisions about the best and most economical ways of managing that waste. The rationale behind introducing SWMP is to ensure that the element of waste generation is thought about right from the design and specification stage and facilitate the selection of the construction methods and materials that would effectively minimise waste generation.

How does a SWMP manage the waste?

The management process through SWMP relies on the PDCA (Plan–Do–Check–Action) cycle – at pre-construction stage it will describe each type of waste expected to be produced by the site, with estimated quantities of the different types of waste and then identifies the waste-management action proposed. Typically this would include reuse, recycling or recovery of materials either on or off the site and, if none of these options are suitable, disposal, i.e. sending to a landfill site. It will also separately distinguish the non-hazardous and hazardous (e.g. asbestos) waste. This planned process will then be put into action at the construction stage by the principal contractor, who would be responsible for implementing the plan, as well as recording actual waste generation and disposal figures against the estimated figures, thus checking the effectiveness of the plan. At the end of the construction phase, it is anticipated that all parties involved, including the client, the designers and the principal contractor will review the SWMP to identify the lessons learned for future projects.

As a basic minimum, a SWMP will contain the following information:

- ownership of the document
- information about who will be removing the waste
- the types of wastes to be removed
- details of the site(s) where the wastes are being taken to

■ A post-completion statement confirming that the SWMP was monitored and updated on a regular basis.

A successful SWMP requires careful planning, preparation and implementation. The bigger the project, the more complex, in depth and time consuming the creation of the SWMP will be.

Who is responsible for SWMP?

The client is responsible for instigating the SWMP, and this must be prepared before construction work starts. The definition of construction work is the same as the definition included in the CDM Regulations.

The designers also have a responsibility to provide adequate information for the SWMP as well as ensuring that due care has been undertaken during the design development in order to minimise waste generation.

Once work starts on site, the principal contractor becomes responsible for managing the SWMP ensuring that it accurately reflects the progress of the project. If the project is started without a SWMP, the client and the principal contractor are both guilty of an offence punishable by a fine of up to £50,000.

What are the benefits of SWMP?

■ Commercial benefit: by encouraging waste minimisation, recycling and reuse.

■ Transparency: by utilising an open-book detailed recording process.

■ Environmental benefit: by reducing and recycling waste.

■ Marketing tool: a well-devised and implemented SWMP will invariably add to corporate success documentation for future reference.

■ Lessons for the future: reviewing SWMP of a completed project will provide good reference material for planning waste management for future projects.

An example of a **SWMP**

Site Waste Management Plan/Checklist				
Project				
Project Location				
Principal Contractor				
Project Manager				
Person responsible for waste management on site				
Person completing form (if different)				

Project Stage		Questions		yes/no	Comments (If 'yes', what action have you taken/do you propose to take? If 'no', why not?)
Policy	1	Has your organisation adopted a waste management policy			
	2	Has client signed Site Waste Management Plan?			
	3	Have relevant sub-contractors producing significant waste streams been identified?			
	4	Have the identified sub-contractors signed Site Waste Management Plan			
Procurement	5	Has an evaluation of materials been made so over-ordering and site wastage is reduced?			
Design	6	Is unwanted packaging to be returned to the supplier for recycling or re-use?			
	7	Is unused materials used on the next project?			
Planning	8	Does Principal Contractor understand their responsibility for waste management planning and compliance with environmental legislation?			
	9	Has project programme been developed to include likely waste arisings (how much, when, and what types)?			
	10	Has an area of the site been designated for waste management, including segregation of waste?			
	11	Have targets been set for waste recycling?			

Part 1 Appendices

		12	Have measures been put in place to deal with expected (and unexpected) hazardous waste?		
		13	Has disposal of liquid wastes been considered?		
		14	Have opportunities been considered for re-use and recycling of materials off site?		
		15	Have you considered what are the most appropriate sites for disposal of residual waste from the project?		
		16	Are there opportunities for reducing disposal cost from waste materials which may have a commercial value?		
	Site Operations	17	Has responsibilities for waste management on site and compliance with environmental legislation been assigned to a named individual?		
		18	Have toolbox talks been planned for all site personnel about waste management?		
		19	Are selected waste materials segreggated on site?		
		20	Are containers/skips clearly labelled?		
		21	Are Duty of Care procedures complied with (ie, transfer notes and checking authorisation of registered carriers, registered exempt sites and licensed waste management facilities?		
		22	Will on site waste management be monitored?		
	Post Completion	23	Has final Report of use of recycled materials, waste reduction, segregation, recovery and disposal, with costs and savings identified been completed?		
		24	Have key waste management issues been considered for action at future projects?		

Comments/Notes

Document Date:

Revision No:

Site waste management plan data sheet

Project:		Principal contractor:		Date:	
Project location:		Project manager:		Date submitted:	
Person responsible for waste management on site:					
Person completing form (if different):					
Types of waste arising (add more rows if needed):					

Material	Quantity (in tonnes)				Total Waste (kg)	Waste Contractor Information				
	Re-used	Recycled (estimated)	Sent to WML-exempt site	Disposal to landfill		Contractor (waste carrier and/or receiver)	Registered waste carrier certificate No.	Expiry date	Waste management licence No.	Expiry date
(List of wastes)										
Inert										
Active										
Hazardous										
Total (tonnes)										
Percentage of Waste Recycled/Reused:										

Part 1 Appendices

APPENDIX 28 Alternative procurement option

Private finance initiative/public–private partnership projects

There are fundamental differences in the managing and delivering of private finance initiative/public–private partnership (PFI/PPP) projects to that of more conventional developments. The key area of difference is that they are set up on a completely different structure and contractual basis. This involves additional parties, a substantially different risk matrix and more complex responsibility for the management of risk. It is important to understand these risks, which are more extensive, more onerous and which extend for a much longer period.

First, however, a brief history of the development of PFI/PPP in the UK Is set out to give an insight of the basis of these changes.

Brief history

In the early 1990s the British government formulated and awarded specific design, build, finance and operate (DBFO) projects to modernise the UK's ageing road infrastructure; these were the forerunners to 'fully' privately financed projects. Also, there were early build, own and operate (BOO) projects and build, own, operate and transfer (BOOT) projects but the DBFO projects were more numerous. These schemes allowed various road infrastructure projects to be built and operated by private companies but more importantly, financed by private capital with the borrowings remaining 'off balance sheet' for the government. However, these private loans were normally 100% underwritten by the state and repaid over a concession period of (initially) 20–25 years through 'shadow tolls' for traffic usage. In effect the government was buying upgraded assets on hire purchase.

Such projects tended to be multimillion-pound developments such as major new motorways and strategic bridge links such as the Second Severn Crossing and the Dartford M25 Toll Bridge.

In the mid-1990s John Major's government then launched PFI. PFI contracts were of a similar format but were used to include renewal and upgrades of other key state-funded facilities. There were two main forms of contract, which passed down different levels of risk to the private sector, namely:

- 'availability' such as schools, prisons, hospitals

- 'full risk' such as water projects, light rail and private roads.

The 'availability' model followed very much the same format as DBFO in that 'availability for use' payments were made, much in line with 'shadow tolls', during the concession period. Where the 'full risk' was passed down, the lengthy concession period allowed revenue to be generated by the private companies directly from the service provided and was used to finance and service their equity and loans. These 'full-risk' projects were not underwritten by the government.

The structure of a PFI/PPP company

In order to allow investment in PFI/PP projects and to permit the effective risk transfer to safeguard the private investors, a new type of legal company had to be created. Special service companies (SPC) or special service vehicles (SPV) were established as the main contracting companies. These SPC/SPVs became not only responsible for the construction of the specific project, but had to operate the facility for up to 30 years before handing it back to the government. A typical structure of a SPC/SPV is shown below:

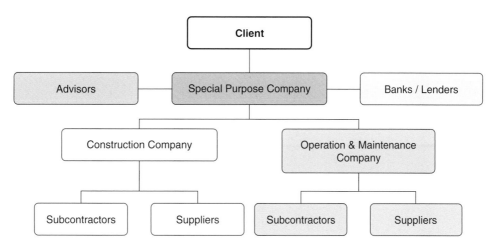

Within this structure there needed to be formal contracts to ensure that all parties understood their legal responsibilities and then were held to account to deliver to the specific requirements and performance criteria. As, by and large, the companies forming the SPC/SPV companies were the same as those involved in the construction and operation and maintenance companies, there needed to be formal arm's-length standalone contracts between the different legal entities.

Project management of PFI/PPP projects

From the diagram above it is obvious that it becomes very important to define exactly which part of the project that a project manager is to be responsible for, namely:

- project manager/director for the SPC/SPV company

- project manager for the construction company or consortium

- project manager for the long-term operating and maintenance company or joint venture.

This is essential as the project manager's role, responsibilities and authority levels will vary greatly. The other critical issue is that the project manager needs to understand fully within which part of the organisation that certain risks should be retained and controlled and then ensure that the appropriate risks are transferred to the legal party contracted to manage them. The diagram below demonstrates this:

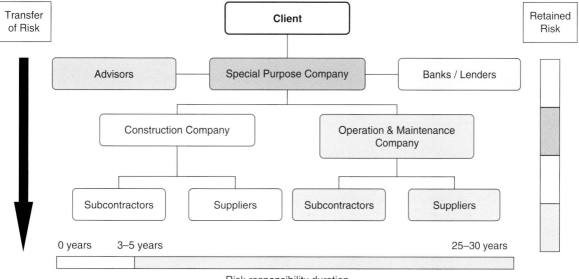

Part 1 Appendices

Risk transfer

At the start of PFI idea, full risks were not fully identified and understood. Thus, in some cases the proper risks were not properly transferred from party to party, hence risks were not always allocated properly to the correct party. Risk transfer has developed over the years and is now a mature market. The best performing PFI/PPP projects have generally been where the risks are properly identified and then allocated to the party best able to understand them and so properly manage them.

APPENDIX 29 Building information modelling

Introduction

Building information modelling (BIM) is the process of generating and managing building data using computer software. The range of information captured through this three-dimensional dynamic and potentially real-time application encompasses building geometry, spatial relationships, geographic information, and quantities and properties of building components thus adding benefits from project inception to occupation stages.

BIM is a process which changes the definition of traditional design development phases and encourages more information sharing than most are used to. This is achieved by modelling representations of the actual parts and elements being used to build a facility. This is a substantial shift from the traditional computer-aided drafting method of drawing (e.g. CAD) with vector-file-based lines that combine to represent objects. While initially developed in the USA, it is gradually getting popular in the UK and the rest of Europe.

Construction documents, including the drawings, procurement details (particularly for estimating purposes), environmental requirements, submittal processes and other specifications require a degree of interoperability between various functions (such as architect, engineer, estimator, constructor, facility management and end user). It is anticipated its proponents that BIM can be used to bridge the knowledge and information loss associated with handing a project from design team to construction team and to facility owner/operator, by allowing each group to add to and reference back all the information they acquire during their period of contribution to the BIM model. For example, a building owner may find evidence of a leak in a building. Rather than exploring the physical building, he may turn to his BIM and see that a water valve is located in the suspect location. He could also have in the model the specific valve size, manufacturer, part number, and any other information ever researched in the past.

Advocates of BIM suggest that BIM offers multiple benefits including improved visualisation, improved productivity due to easy retrieval of information, increased co-ordination of construction documents, embedding and linking of vital information such as suppliers for specific materials, location of details and quantities required for estimation and tendering, increased speed of delivery and overall reduced costs. Technologies similar to BIM include Virtual Building Environment (VBE), Virtual Building, BuildingSMART, Integrated Practice and Virtual Design and Construction (VDC).

BIM is certainly viable and offers many realisable advantages over CAD; however, the ability to share intelligent building information is perhaps its most critical advantage.

The use of BIM for improved management of construction projects

Lean construction thinking applied to construction production systems has increased the awareness of the benefits of stable work, of the pull-flow of teams and materials to reduce inventories of work in progress, and of process transparency to all involved. Three-dimensional visualisations of process status and future direction, delivered to all on site, are either essential or at least highly beneficial for all of these. They can empower people working on the site to manage the day-to-day flow of construction operations with less direct control from higher levels of management, with better quality and less waste. However, it is very difficult to visualise the flow of the work in progress on a construction site. This is obvious for the interior finishing works that comprise the majority of the value in most buildings, but in terms of production flow, it is also true for structural work: the amount of buffered work-in-progress accumulated between work teams cannot be seen by the naked eye. Therefore, in construction, forming a coherent view of a project's flow status requires integrating and interpreting monitoring data gathered from various sources. Computer-aided visualisation, not only of the construction product, but also of the construction process, can facilitate reporting of project status. More importantly, however, it can provide a unique service to support decision-making to achieve stable flows and to communicate pull-flow signals. BIM is an approach to designing, simulating, detailing and managing construction in which a machine-readable parametric data model of the facility is compiled and used by the project team using appropriate computer software. Construction can then be managed through BIM-based tools to deliver construction process information directly to all personnel in a construction project, and particularly to those working on the site, in ways that empower them to visualise the state of the process and control their role in it. Such software applications enable leveraging BIM information in order to improve the flow of materials, people, equipment and information on construction sites.

APPENDIX 30 Business case development

A business case is used to obtain management commitment and approval for funding or investment through demonstrating a rationale for the funding or investment. The business case provides a framework for planning and management of the proposed development as well as a monitoring benchmark for the ongoing viability of a project.

A template is provided below for the development of a business case.

1.   Definition of the project proposal.

2.   Objective of the project proposal.

3.   Strategic fit.

    3.1   Business need.

    3.2   Organisational overview.

    3.3   Contribution to key organisational objectives.

    3.4   Stakeholders.

    3.5   Existing arrangements.

    3.6   Scope (minimum, desirable and optional).

    3.7   Constraints.

    3.8   Dependencies.

    3.9   Strategic benefits.

    3.10 Strategic risks.

    3.11 Critical success factors.

4.   Options appraisal.

    4.1   Long and short list of options.

    4.2   Opportunities for innovation and collaboration.

    4.3   Service delivery options – who will deliver the project?

    4.4   Environmental, social and economic criteria.

    4.5   Implementation options.

    4.6   Detailed options appraisal demonstrating value for money and sustainability.

    4.7   Risk quantification and sensitivity analysis.

    4.8   Benefits appraisal.

    4.9   Preferred option.

5.   Commercial aspects.

    5.1   Output-based specification.

    5.2   Sourcing options.

    5.3   Risk allocation and transfer.

    5.4   Contract length.

    5.5   Implementation timescales.

6.   Affordability.

     6.1   Budgetary issues.

     6.2   Income and expenditure.

     6.3   Cashflow prediction.

7.   Achievability.

     7.1   Evidence of similar projects.

     7.2   Project roles.

     7.3   Delivery strategy.

     7.4   Risk management strategy.

     7.5   Benefits realisation plan.

     7.6   Contingency plan

The effectiveness of a business plan should be judged on the basis of the following key criteria:

■   Is the need for the project clearly stated?

■   Have the benefits been clearly identified?

■   Are the reasons for and benefits of the project consistent with the overall strategy?

■   Is it clear what will define a successful outcome?

■   Is it clear what the preferred option is?

■   Is it clear why this is the preferred option?

■   Is it clear what the procurement option is?

■   Is it clear why this is the preferred procurement option?

■   Is it clear how the necessary funding will be put in place?

■   Is it clear how the benefits will be realised?

■   Are the risks faced by the project explicitly stated?

■   Are the plans for addressing those risks explicitly stated?

(Source: information distilled from OGC publications.)

APPENDIX 31 Key sustainability issues

Table A Sustainability issues – design and objectives (source: *CIBSE Introduction to Sustainability*. © April 2007 The Chartered Institution of Building Services Engineers, London)

Sustainability issue	Examples of design and construction objectives
Energy and carbon dioxide	Reduce predicted carbon dioxide emissions by applying energy-efficient design principles and using low- and zero-carbon technologies
Water	Reduce predicted water use by integrating water-efficient plant, appliances and fittings
Waste	Reduce construction and demolition waste going to landfill and enable in-use recycling in accordance with the waste hierarchy
Transport	Increase the use of sustainable modes of transport when the building is in use
Adapting to climate change	Improve the capacity of the building to operate successfully under the different and demanding conditions predicted in future
Flood risk	Mitigate the risk of flooding (and design for flood resilience)
Materials and equipment	Reduce the embodied lifetime environmental impact by selecting on the basis of environmental preference
Pollution	Reduce unavoidable building-related emissions and the risk of accidental pollution
Ecology and biodiversity	Enhance the ecology and biodiversity of the site by protecting existing assets and by introducing new habitats and/or species
Health and well-being	Provide a safer, more accessible, healthy and comfortable environment
Social issues	Reduce crime and adverse effects on neighbours throughout the lifetime of the development through design and good practice in construction and operation

Table B Key actions at each project stage (source: *CIBSE Introduction to Sustainability*. © April 2007 The Chartered Institution of Building Services Engineers, London)

Key stage	Key actions
Pre-inception	Identify all drivers for sustainability and ensure that appointment allows for project team to respond to these drivers
	Identify the risks associated with project which relate to sustainability (e.g. flood risk assessment, damage to ecological habitat, transport impacts, etc.)
	Determine potential impact of sustainability targets (e.g. a target for a 'zero-carbon development' is likely to have implications on whole project team)
	Include scope and fees for early-stage predictions of energy and water use in scope of work (early-stage energy/carbon assessments are becoming essential)
	Determine whether an environmental impact assessment is required
Strategic brief	Provide a response to the strategic brief by considering drivers for sustainability and raising issues early in the project
	Identify any requirements in the brief that could conflict with sustainability objectives (e.g. design targets for low internal temperatures in summer)
	Identify requirements for input from specialist consultants (e.g. likely ground conditions for ground source heat pumps)
Project brief	Propose sustainability objectives and targets, in particular carbon and water targets in response to drivers for sustainability
	Determine whether assessment methodologies are required (e.g. BREEAM, NEAT) and ensure that project contributes towards all relevant targets

Key stage	Key actions
Strategy	Ensure that design responsibilities are allocated for all critical sustainability targets, especially those relating to carbon and water use
	Undertake an initial site analysis against sustainability targets, including determining infrastructure capacity, establishing ground conditions, etc.
	Provide rules of thumb and design guidance for project team on key issues (e.g. number of wind turbines required to meet predicted loads, or likely spaces for an energy centre)
	Develop an energy and carbon emissions strategy by following the principles set out in CIBSE *Guide L: Sustainability*
	Develop a water management strategy by following the principles set out in CIBSE *Guide L: Sustainability*
	Develop a strategy for adapting to the effects of climate change by following the principles set out in CIBSE *Guide L: Sustainability*
	Recommend that the project team establishes the flood risk of the site and consults with local authority to determine whether a strategic flood-risk assessment has been undertaken
	Incorporate flood-resistant principles into the design of building services and work with the design team to raise awareness of flood risk and flood resistance
	Recommend that project teams give consideration to the incorporation of sustainable drainage systems and the potential to integrate with rainwater collection
	The project team should liaise with transport planners, in order to identify the scope of transportation work required by the local authority
	Recommend that a suitably qualified ecologist be involved to undertake and ecological appraisal of the site
	Inform the project team of shading benefits of vegetation integrated into the building design and landscape (e.g. green roofs or walls)
	Incorporate access and inclusion measures identified in the accessibility audit
	Recommend that a waste management strategy be prepared for the operation of the building
	Consider potential for energy from waste systems
	Establish the need for and feasibility of waste management facilities such as compactors, serviced storage spaces, etc.
	Recommend that the lifecycle impacts of materials and equipment are considered by the project team and that these are considered during the selection of construction methods in terms of ventilation strategies, appropriate thermal mass, etc.
	Make the project team aware of the principles of designing for deconstruction and consider the whole life of services components for recycling or reuse at the end of their life
	Recommend that there is active engagement and consultation with the local community
	Highlight the need for consultation with the local police architectural liaison officers on safety and security ('Secure by Design')
	Determine the planning strategy and establish the information that is required for the submission. In particular, determine whether an energy strategy report and sustainability statement are required for application
	Contribute towards an environmental impact assessment (if required), particularly in relation to air quality, noise, microclimate issues, etc.
Design	Identify the options for reducing demand, supplying efficiently and for providing low- or zero-carbon technologies
	Propose feasible technologies and techniques to meet carbon emissions targets

Key stage	Key actions
Construction	Identify the options for reducing water demand, supplying water efficiently and for use of rainwater or treatment and reuse of water
	Propose feasible technologies and techniques to meet water-use targets
	Advise clients on the maintenance and operational implications associated with using F-Gas refrigerants, such as R134a and R407c.5.3
	Ensure that the proposals provide comfortable and appropriate internal conditions that promote health and well-being, as set out in the relevant guidance
	Ensure that storage space for efficient management of waste and recyclable material during operation is incorporated into the layout and that this space is correctly serviced and managed
	Select and source materials based on the overall environmental impacts and suppliers' declarations
	Avoid use of environmentally hazardous materials such as insulants with gases implicated in global warming
	Avoid selecting or locating plant that may create additional noise over the existing background level
	Consider alternative arrangements for providing and maintaining infrastructure and delivering services, such as energy service companies and multi-utility joint ventures
	Incorporate all technologies and techniques, as identified in the earlier design stages and refer to the CIBSE online sustainable engineering tool to identify detailed measures
	Recommend that contractor selection takes account of environmental credentials
	All relevant tender packages should be reviewed against the sustainability requirements for the project
	Recommend that subcontractor and supplier selection takes account of environmental credentials
	Recommend a periodic review of sustainability performance against objectives and targets
	Ensure that the engineering services that are procured and delivered to site meet the performance standards relating to sustainability, and that the requirements are fully addressed
	Observe construction site practices and comment on practices that could have a significant impact on the environment
Commissioning	Ensure that systems building commissioning/recommissioning results accord with sustainability targets and that the contractor is notified of any issues with performance
Building handover	Provide a building logbook and occupant user guide for projects and ensure that there is a clear explanation of design targets and assumptions to allow comparison with actual, operational energy use
Operation	Ensure that the system is operating according to design intent, which may involve periodic recommissioning and post-occupancy evaluation
	Recommend that sustainability be addressed when the building owner or occupier specifies tenders and evaluates contracts for the operation and maintenance of facilities
	Ensure that refurbishment or refit projects implement the relevant sustainability principles, as set out in this document
	Undertake energy and water management activities including audits and benchmarking to identify potential for further savings
	Recommend that projects to re-engineer systems consider potential reuse of materials or systems

Key stage	Key actions
Deconstruction	Ensure that audits and condition surveys include assessment against key sustainability drivers and targets, as identified in this document
	Refer to the CIBSE online sustainable engineering tool to identify detailed measures for improving performance
	An audit should be undertaken prior to demolition commencing to identify the potential for cost-effective recovery of material from demolition

APPENDIX 32 Environmental sustainability assessment methods

BREEAM:

The Building Research Establishment Assessment Method (BREEAM 2008) is the leading and most widely used environmental assessment method for buildings (other than homes and dwellings). It sets the standard for best practice is sustainable design and grades performance as pass, good, very good, excellent and outstanding. The assessment of the environmental impacts is carried out at:

■ Design stage: leading to an interim BREEAM certificate.

■ Post-construction stage leading to a final BREEAM certificate and is measured against the following 10 categories:

- ○ management
- ○ health and well-being
- ○ energy
- ○ transport
- ○ water
- ○ material
- ○ waste
- ○ land use and ecology
- ○ pollution
- ○ innovation.

The table below summarises the main issue in each of the BREEAM categories.

Categories	Issues
Management	Commissioning
	Construction site impacts
	Security
Health and well-being	Daylight
	Occupant thermal comfort
	Acoustics
	Indoor air and water quality
	Lighting
Energy	Carbon dioxide emissions
	Low- or zero-carbon technologies
	Energy sub-metering
	Energy efficient building systems
Transport	Public transport network connectivity
	Pedestrian and cyclist facilities
	Access to amenities
	Travel plans and information
Water	Water consumption
	Leak detection
	Water reuse and recycling

Part 1 Appendices

Materials	Embodied lifecycle impact of materials	
	Materials reuse	
	Responsible sourcing	
	Robustness	
Waste	Construction waste	
	Recycled aggregates	
	Recycling facilities	
Land use and ecology	Site selection	
	Protection of ecological features	
	Mitigation/enhancement of ecological value	
Pollution	Refrigerant use and leakage	
	Flood risk	
	Nitrous oxide emissions	
	Watercourse pollution	
	External light and noise pollution	
Innovation	Exemplary performance levels	
	Use of BREEAM accredited professionals	
	New technologies and building processes	

Home and dwellings

Code for sustainable homes

For homes and dwellings there are two assessment methods:

- *Code for Sustainable Homes*, April 2007 (new housing in England) issued by the Department for Communiities and Local Government.

- The Building Research Establishment's *EcoHomes*, 2006 (existing homes in England and all homes in Scotland, Wales and Northern Ireland). This assessment method formed the basis for the above code.

The *Code for Sustainable Homes* measures sustainability against nine design categories:

- energy and carbon dioxide emissions

- water

- materials

- surface water run-off

- waste

- pollution

- health and well-being

- management

- ecology.

The table below summarises the main issues in each of the code categories:

Categories	Issues
Energy and carbon dioxide emissions	Dwelling emission rate (M) Building fabric Internal lighting Drying space Energy labelled white goods External lighting Low- or zero-carbon technologies Cycle storage Home office
Water	Internal water use (M) External water use
Materials	Environmental impact of materials Responsible sourcing of materials – building elements Responsible sourcing of materials – finishing elements
Surface water run-off	Management of surface water run-off from developments (M) Flood risk
Waste	Storage of non-recyclable waste and recyclable household waste (M) Construction waste management (M) Composting
Pollution	Global warming potential of insulants Nitrous oxide emissions
Health and well-being	Daylighting Sound insulation Private space Lifetime homes (M)
Management	Home use guide Considerate constructors scheme Construction site impacts Security
Ecology	Ecological value of site Ecological enhancement Protection of ecological features Change in ecological value of site Building footprint
(M) denotes mandatory elements.	

The code uses a rating system of one to six stars with one star being the entry level just above the Building Regulations and six stars being evidence of exemplar sustainable development. Credits are given for achieving a certain level of performance in each category and the factored against a weighting system see the figures below.

Scoring system for the *Code for Sustainable Homes*

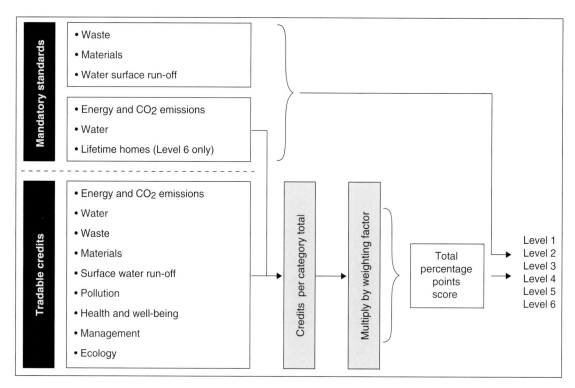

Calculating the total points score

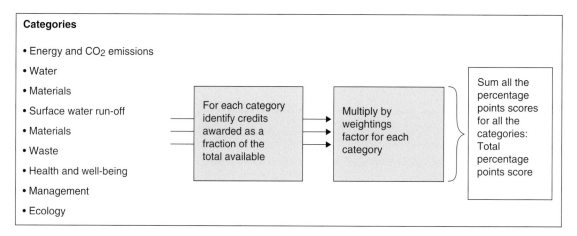

As with BREEAM, the assessment is carried out at the design and post-construction stages.

EcoHomes

EcoHomes measures sustainability against eight similar design categories:

■ energy

■ transport

■ pollution

■ materials

■ water

■ land use and ecology

- health and well-being

- management

and employs a similar credit rating system which classifies the development under pass, good, very good and excellent. The table below summarises the main issues in each EcoHomes category.

Categories	Issues
Energy	Dwelling emission rate
	Building fabric
	Drying space
	EcoLabelled goods
	Internal lighting
	External lighting
Transport	Public transport
	Cycle storage
	Local amenities
	Home office
Pollution	Insulant global warming potential
	Nitrous oxide emissions
	Reduction and surface run-off
	Renewable and low emission energy source
	Flood risk
Materials	Environmental impact of materials
	Responsible sourcing of materials: basic building elements
	Responsible sourcing of materials: Finishing elements
	Recycling facilities
Water	Internal potable water use
	External potable water use
Land use and ecology	Ecological value of site
	Ecological enhancement
	Protection of ecological features
	Change of ecological value of site
	Building footprint
Health and well-being	Daylighting
	Sound insulation
	Private space
Management	Home use guide
	Considerate constructors
	Construction site impacts

Part 1 Appendices

Part 2
Project handbook

Project handbook

Introduction

The purpose of the handbook is to guide the project team in the performance of its duties, which are the design, construction and completion of a project to the required specifications within the approved parameters of the contract budget and to schedule. In practice, a project handbook should be concise, clear and consistent with all other contract documentation and terms of engagement. The emphasis should be to identify policies, strategies and the lines of communication and key interfaces between the various parties. It is important that the handbook is tailored to fit the needs of each project. The comprehensive format given here would be too bulky for some projects with the danger of it being ignored.

The handbook is prepared by the project manager in consultation with the project team where possible at the beginning of the pre-construction stage and describes the general procedures to be adopted by the client and the team. It comprises a set of ground rules for the project team. It differs from the project execution plan, which is primarily written for the client and funding partners, in giving a route map through the stages and processes of the project demonstrating financial control and a *modus operandum* to achieve the project objectives.

The handbook is not a static document and it is anticipated that changes and amendments will be required in accordance with procedures as later outlined. Consequently, a loose-leaf format should be adopted to facilitate its updating by the project manager who is the only person authorised to co-ordinate and implement revisions. Copies of the handbook will be provided to each nominated member of the project team as listed under parties to the project.

Aims of the handbook

The handbook's aim is to identify responsibilities and co-ordinate the various actions and procedures from other documents/data already or currently or likely to be prepared into one authoritative document covering as a rule and depending on the nature/scope of project, the main elements and activities outlined in the following sections.

Parties to the project

This section will include the following items:

- A list of all parties involved in the project including those employed by the client as well as their contact details (addresses, phone and fax numbers and email address).

- The name of the project manager responsible for the project together with details of his duties responsibilities and authority (see Appendix 1 of Part 1).

- Details of other team members and/or stakeholders involved complete with their duties, responsibilities and contact details.

- Organisational charts indicating line and functional relationships, contractual and communication links and any changes to suit the various stages/phases of the project.

Third parties

This section will provide the names and contact details of all legal authority departments, public utilities, hospitals, doctors, police stations, fire brigade, trade associations, adjoining landowners, adjacent tenants and any other bodies or persons likely to be involved.

Roles and duties of the project team

The information provided should be the minimum necessary to facilitate the understanding of the roles of the others involved by each member of the team. The services to be provided are described by reference to standard agreements or contracts with any amendments and additions included. The aim is to ensure that there are no gaps or overlaps.

Project site

Details will be provided of prevailing relevance of arrangements for demolition, clearing and diversion of existing services, hoardings and protection to adjacent areas (e.g. noise pollution).

General administration including communication and document control

The project manager will be responsible for the following items:

- The adequacy of all aspects of project resourcing (staffing, equipment and aids, site offices and welfare accommodation).

- Office operating systems and routines so that the staff know them and are applied consistently and efficiently.

- Providing suitable working accommodation and facilities for members of the project team and for meetings or group discussions.

Action will need to be taken by the project manager in respect of documentation control, storage, location and retrieval; this will affect:

- Letters, contract documents, reports, drawings, specifications, schedules, including financial and all specialist fields (e.g. facilities management, technology, health and safety, environmental).

- Accessibility for updating.

- Records for all documents/files and control of their movement.

- Office security: (1) storage of legal documents (originals and duplicates); (2) entry safeguard, fire and intruder alarms.

Part 2 Project handbook

■ Retention of documents/files on project completion/suspension: (1) archive storage – legal and contractual time limits; (2) dead files – removals and destruction and their register.

All correspondence should be headed by the project title and identified by:

■ Subject/reference of communication.

■ Addressee's full details.

■ Those parties receiving copies.

Each piece of correspondence should refer to a single matter or a series of direct and closely linked matters only. Distribution of copies should be decided on the basis of the subject matter and confidentiality against a predetermined list of recipients. All communications between the parties of the project involving instructions must be given in writing and the recipients should also confirm it in the same manner.

Contract administration

Contract conditions

It is essential that there is an understanding of the terms of all contracts and their interpretation by all concerned. The role of parties, their contribution and responsibilities, including relevance of timescales and client–project manager operating and approvals pattern will have to be established.

Contract management and procedures

Matters associated with contract management will include forms of contract for contractors/subcontractors; works carried out under separate direct contract; procedures for the selection and appointment of contractors/subcontractors; checklists for design team members and consultants meeting their supervisory and contractual obligations (e.g. inspections and certification); the placement of orders for long delivery components and the preparation of contract documents.

Tender documentation

Design and specifications details to be included; tender analysis and reporting; lists of tenderers and interview procedures; system for the preparation of documents and their checking; award and signature arrangements.

Assessment and management of variations

(See Appendix 13 of Part 1 for the change management checklist and Appendix 16 for the dispute resolution processes.)

Extensions of time

■ The project manager has the responsibility for ensuring that there is early warning, hence creating the possibility of alternative action/methods to prevent delay and additional costs.

■ A schedule should be prepared, stating the grounds for extension, relevant contract clauses and forecast of likely delay and cost.

■ The involvement and possible contribution to the solution of problems of other parties affected should be established.

■ A procedure will need to be available for extension approval. If relevant, the disputes procedure may be invoked.

Loss and expense

■ Applicable procedures are covered under standard or in-house forms of contract relevant to the specific project.

Indemnities, insurances and warranties

Relevant provisions depend on the nature of the project. However, they are usually governed by the conditions specified by the forms of 'model' contracts/agreements issued by professional bodies or those in common use in the construction industry. Typical examples of insurances applicable to construction projects include:

■ Contractors' all-risk policies (CAR), usually covering loss or damage to the works and the materials for incorporation into the works; the contractor's plant and equipment including temporary site accommodation; the contractor's personal property and that of employees (e.g. tools and equipment). The CAR policy is normally taken out by the contractor but should insure in the joint names of the contractor and the client (employer). Subcontractors may or may not be jointly insured under the CAR policy.

■ Public liability policy – this insures the contractor against the legal liability to pay damages or compensation or other costs to anyone that suffers death, bodily injury or other loss or damage to their property by the activities of the contractor.

■ Employers' liability policy – every contractor will have this either on a company-wide basis, covering both staff and labour, or on a separate basis for the head office and for each site separately.

■ Professional indemnity (PI) – the purpose of this is to cover the liabilities arising out of 'duty of care'. Typically, consultants (including the project manager) will require this policy to cover their design or similar liabilities and liabilities for negligence in undertaking supervision duties. In case of a design and build contract, the contractor has to take out a separate PI policy, as designing is not covered by the normal CAR policy.

Design co-ordination

The project brief will be reviewed jointly by client and project manager with the aim of confirming that all relevant issues have been considered. These may include

■ health and safety obligations

■ environmental requirements

■ loading considerations

■ space and special accommodation requirements ⎫ Eventual user's

■ standards and schedule of finishes ⎭ needs

■ site investigation information/data

■ availability of necessary surveys and reports

■ planning consent and statutory approvals

■ details of internal and external constraints.

The project manager will need to seek the client's approval to issue the brief and relevant information to the design team and other consultants. Among the other duties which fall to the project manager are the following:

- Defining the roles and duties of the project team members.

- Responsibility for the drawings and specifications:

  - establishing format (e.g. CAD compatibility issues) sizes and distribution and seeking comments on their content and their timing

  - the issue of tender drawings and specifications

  - advising contractors and subcontractors of the implications of the design

  - setting requirements for: (1) shop/fabrication drawings; (2) test data and (3) samples and mock-ups.

- Monitoring the production of the outline proposals for the project by:

  - reviewing sketch plans and outline specifications in terms of the brief

  - preparing the capital budget and reconciling this with the outline budget

  - appraising the implications of the schedule

  - effecting reconciliation with the project master schedule

  - finalising the outline proposals making recommendations/presentations to the client and seeking the latter's approval to proceed.

- Monitoring design work at the pre-tender stage by:

  - reviewing with the consultants concerned the client's requirements, brief documentation and their sectional implications

  - agreeing team members' input and identification of items needing client's clarification

  - reviewing with the client any discrepancies omissions and misunderstandings, seeking their resolution and confirming to the team.

- Agreement of overall design schedule and related controls.

- Identification of items for pre-ordering and long delivery preparation of tender documents, client's approvals and placement of orders and their confirmation.

- Monitoring production of drawings and specifications throughout the various stages of the project and their release to parties concerned.

- Arranging presentations to the client at appropriate stages of design development and securing final approval of tender.

Change management design

- Reviewing with the design team and other consultants any necessary modifications to the design schedule and information required schedules (IRS) in the light of the appointed contractors'/subcontractors' requirements and reissuing revised schedule/IRS.

Part 2 Project handbook

- Preparing detailed and specialist designs and subcontract packages including bills of quantities.

- Making provision for adequate, safe and orderly storage of all drawings, specifications and schedules including the setting up of an effective register/records and retrieval system.

The project manager must ensure that the client is fully aware that supplementary decisions must be obtained as the design stages progress and well within the specified (latest) dates in order to avoid additional costs. Designs and specifications meeting the client's brief and requirements are appraised by the quantity surveyor for costs and are confirmed to be within the budgetary provisions.

Handling changes will require a series of actions. The project manager will be responsible for these activities:

- Administering all requests through the change order system (see Table 5.3 and Appendix 17 for checklist and specimen form).

- Retaining all relevant documentation.

- Producing a schedule of approved and pending orders which will be issued monthly.

- Ensuring that no changes are acted on unless formally decided.

- Considering amendments and alterations to the schedules and drawings within the provisions of the applicable contract/agreement.

- Initial assessment of any itemised request for change made by the client taking due account of the effect on time.

Action by consultants in relation to variations will include the following items:

- Securing required statutory/planning approvals and cost-checking revised proposals. Confirmation of action taken to project manager.

- Design process and preparation of instructions to contractors involved.

- Cost agreement procedure for omissions and additions, i.e. estimates, disruptive costs, negotiations and time implications.

Site instructions

Site instructions must be issued in writing and confirmed in a similar manner by recipients. Site instructions which constitute variations can be categorised as:

- normal

- special (e.g. concerned with immediate implementation as essential for safety, health and environmental protection aspects)

- extension of time required or predicted

- additional payments involved or their estimate.

Site instructions will be binding if they are issued and approved in accordance with the contract provisions.

Cost control and reporting

The quantity surveyor has overall responsibility for cost monitoring and reporting with the assistance of and input from the design team, other consultants and contractors. Action at the pre-construction stage involves the following items:

■ The preparation of preliminary comparison budget estimates.

■ The agreement of the control budget with the project manager.

■ Project budget being prepared in elemental form; the influence of grants is identified.

■ The establishment of work packages and their cost budgets.

■ Costing of change orders.

Other elements associated with work control are as follows:

■ Assessment of cost implications for all designs, including cost comparison of alternative design solutions.

■ Value analysis procedures, including cost in use.

■ Comparison of alternative forms of construction using data on their methodology and costs.

■ Comparison of cost budgets and tenderers' prices at subcontract tender assessment.

■ Tenders which are outside the budget and which require an input from the project manager on such matters as:

   ○ alteration of specifications to reduce costs

   ○ acceptance of tender figure and accommodating increased cost from contingency; alternatively the client may accept the increase and seek savings from other areas

   ○ possible retendering by alternative contractors.

■ Production of monthly cost reports, including:

   ○ variations since last report, incorporating reasons for costs increase/ decrease

   ○ current projected total cost for the project

   ○ cash flow for the project: (1) forecast of expenditure; (2) actual cash flow as schedule monitoring device indicating potential overspending and any areas of delay or likely problems.

The report should be agreed with and issued to the project manager who will:

■ give advice and initiate action on any problems that are identified

■ arrange distribution of copies according to a predetermined list.

Planning schedules and progress reporting

Planning is a key area and can have a significant effect on the outcome of a project. The handbook will set out the composition and duties of the planning support team and the appropriate techniques to be used (e.g. bar charts, networks). The planning and scheduling will then follow the steps set out in Appendix 3:

- Preparation of an outline project schedule, which will include co-ordination of design team contractors and client's activities then seeking of the client's acceptance.

- Production of an outline construction schedule indicating likely project duration and the basis for determining the procurement schedule.

- Production of an outline procurement schedule including the latest date for placement of orders (materials equipment contractors) and design release dates.

- Modifications if necessary to the outline construction schedule due to constraints.

- Production of the outline design schedule including necessary modifications due to external limitations.

- Preparation of the project master schedule.

- Preparation of a short-term schedule for the pre-construction stage; this will be reviewed monthly.

- Production of a detailed design schedule in consultation with and incorporating design elements from the design team members concerned including:

  ○ scheme design schedule

  ○ drawing control schedule

  ○ client decision schedule

  ○ agreement by client consultants and project manager

- Reviewing the outline procurement schedule and its translation into one, which is detailed.

- Preparation of a works package schedule.

- Production of schedules for bills of quantities procurement including identification of construction phases for tender documentation and production of tender documentation control.

- Expansion of the outline construction schedule into one, which is detailed.

- Preparation of schedules for:

  ○ enabling works

  ○ fitting out (if part of the project)

  ○ completion and handover

  ○ occupation/migration (if part of the project).

Progress monitoring and reporting procedures should be on a monthly basis and agreed following consultation with consultants and contractors. Reports will need to be supplied to the project manager who will report to the client.

Meetings

Meetings are required to maintain effective communications between the project manager, project team and the other parties concerned, e.g. those responsible for industrial relations and emergencies as well as the client. The frequency and location of meetings and those taking part will be the responsibility of the project manager. Meetings held too frequently can lead to a waste of time whereas communications can suffer where meetings are infrequent. Appendix B contains details of typical meetings and their objectives.

Procedures for meetings include:

■ agenda – issued in advance stating action/submissions required

■ minutes and circulation list (time limits involved); Appendix 14 contains examples of agenda and minutes

■ written confirmation and acknowledgement of instructions given at meetings (time limit involved)

■ reports/materials tabled at meetings to be sent in advance to the chair.

Selection and appointment of contractors

The project manager as the client's representative has the responsibility with the support of relevant consultants for the selection and appointment of:

■ contractors, e.g. main, management, design and build

■ contractors, e.g. specialist, works, trade.

The various processes associated with this activity are summarised below:

■ Selection panel appointments relevant to the nature and scope of tender to be awarded. Nomination of a co-ordinator (contact) for all matters concerned with the tender.

■ Establishment of selection/appointment procedures for each stage.

Pre-tender

Pre-tender activities will include the following:

■ Assessment of essential criteria/expertise required for a specific tender.

■ Preparation of long (provisional) list embracing known and prospective tenderers.

■ Checks against database available to project manager, especially financial viability and quality of past and current work; possible use of telephone questionnaire to obtain additional data.

■ Potential tenderers invited to complete/submit selection questionnaire; short list finalised accordingly.

■ Arrangements for pre-qualification interview including prior issue of the following documentation relevant to the project to the prospective tenderer with interview agenda outline of special requirements and expected attendees to cover:

　○ general scope of contract works and summary of conditions

　○ preliminary drawings and specifications

　○ summary of project master and construction schedules

　○ pricing schedule

　○ safety, health and environmental protection statement

○   labour relations statement

○   quality management outline

■   Tender and reserve lists finalised.

Tendering process

The tendering process includes the following activities:

■   Selected tenderers confirm willingness to submit *bona fide* tenders. Reserve list is employed in the event of any withdrawals and selection made in accordance with placement order.

■   Tender documents issued and consideration given by both parties to whether mid-tender interview is required or would be beneficial.

■   Interview arranged and agenda issued.

Carry out the following on the receipt of all tenders:

■   evaluation of received tenders

■   arrangements for post-tender interview and prior issue of agenda

■   final evaluation and report

■   pre-order check and approval to place order.

Safety, health and environmental protection

The handbook should draw attention to the specific and onerous duties of the client and other project team members under the CDM Regulations, and should include procedures to ensure they cannot be overlooked. It is the responsibility of the principal contractor to formulate the health and safety plan for the project to be adhered to by all contractors in accordance with the CDM Regulations and taking account of other applicable legislation. Contractors are required as part of their tender submission to provide copies of their safety policy statement which outlines safe working methods that conform to the CDM Regulations.

Other matters which come within the remit of the principal contractor are as follows:

■   The establishment and enforcement within the contractual provisions of rules, regulations and practices to prevent accidents, incidents or events resulting in injury or fatality to any person on the site, or damage or destruction to property, equipment and materials of the site or neighbouring owners/occupiers.

■   Arranging first-aid facilities, warning signals and possible evacuation as well as the display of relevant notices posters and instructions.

■   Instituting procedures for:

○   regular inspections and spot checks

○   reporting to the project manager (with copies to any consultants concerned) on any non-compliance and the corrective or preventive action taken

○   hazardous situations necessitating work stoppage and in extreme cases closedown of the site.

Part 2 Project handbook

Quality assurance: outline

This is applicable only if quality assurance (QA) is operated as part of contractual provisions. It is critical for the client to understand the operation of a QA scheme, its application and limits of assurance and the need for defects insurance. Procedures and controls will need to be established to ascertain compliance with design and specifications and to confirm that standards of work and materials quality have been attained. The consultants will review details of their quality control with the project manager. The contractors' quality plan will indicate how the quality process is to be managed, including control arrangements for subcontractors.

Responsibility for monitoring site operation of QA administration and control procedures for the relevant documents will need to be established.

As an alternative to QA, any procedures for the management of quality should be included in the handbook (see under 'Management of quality' in Chapter 5).

Disputes

Procedures for all parties involved in the project in the event of disagreement and disputes are to be specified in accordance with the contractual conditions/ provisions which are applicable.

Signing off

Any procedures for signing-off documents should be specified. Signing-off points may occur progressively during stages of the project and be incorporated in a 'milestone schedule'. Details should include permitted signatories and a distribution list.

Reporting

The following reports are examples of what might be prepared.

Project manager's progress report

To be issued monthly and include details of:

- project status
  - updated capital budget
  - accommodation schedule
  - authorised change orders during the month
  - other relevant matters
- operational brief
- design development status
- cost plan status and summary of financial report
- schedule and progress:
  - design
  - construction
- change/variation orders

Code of Practice for Project Management

- client decisions and information requirements
- legal and estates
- facilities management
- fitting out and occupation/(migration) planning
- risks and uncertainties
- update of anticipated final completion date
- distribution list.

Consultant's report

Issued monthly and including input from consultants and containing details on:

- design development status
- status of tender documents
- information produced during the month
- change orders/design progress
- information requirements/requests status
- status of contractor/subcontractor drawings/submittals
- quality control
- distribution list.

Financial control (QS) report

Issued monthly and including:

- reconciliation capital sanction/capital budget
- updated cost plan and anticipated final cost projection
- authorised change orders – effects
- pending change orders – implications
- contingency sum
- cash flow
- VAT
- distribution list.

Daily/weekly diary

Prepared by each senior member of the project team and filed in its own separate loose-leaf binder for quick reference and convenient follow-up. Diaries are made accessible to the project manager and typically contain:

- a summary of forward and *ad hoc* meetings and people attending
- a summary of critical telephone conversations/messages
- documents received or issued
- problems comments or special situations and their resolution

Part 2 Project handbook

234

- schedule status (e.g. work package progress or delays)
- critical events and work observations
- critical instructions given or required
- requests for decisions or actions to be taken
- an approximate time of day for each entry
- a distribution list.

Construction stage

The handbook will include procedures for the following activities:

- Issuing drawings, specifications and relevant certificates to contractors.
- Actioning the consultants' instructions, lists, schedules and valuations.
- Aspects prior to commencement such as:
  - recording existing site conditions, including adjacent properties
  - ensuring that all relevant contracts are in place and that all applicable conditions have been met
  - confirming that all risk insurance for site and adjacent properties is in force
  - ensuring that all site facilities are to the required standard including provisions for health, safety and environmental protection.
- Control of construction work including:
  - reviewing a contractor's preliminary schedule against the master schedule and agreeing adjustments
  - ensuring checks by the main contractor on subcontractor schedules
  - checking and monitoring for all contractors the adequacy of their planned and actual resources to achieve the schedule
  - approvals for subletting in accordance with contractual provisions
  - reporting on and adjusting schedules as appropriate
  - checks for early identification of actual or potential problems (seeking client's agreement to solutions of significant problems).
- Controls for variations and changes (see Appendix 13).
- Controls for the preparation and issue of change orders (see Figure 5.1 and Appendix 13).
- Processing the following applications for the client's action:
  - interim payments from consultants and contractors
  - final accounts from consultants
  - final accounts from contractors subject to receipt of relevant certification
  - payment of other invoices.

- Making contact and keeping informed the various authorities concerned to facilitate final approvals.

- The design team and other relevant consultants to supervise and inspect works in accordance with contractual provisions/conditions and participate in and contribute to:

  o the monitoring and adjustment of the master schedule

  o controls for variations and claims

  o identification and solutions of actual or potential problems

  o subletting approvals

  o preparation of change orders.

Operating and maintenance

The procedures for fitting out should be designed to avoid divided responsibility in the case of failure of parts of the building or its services systems. The procedures to be used in the handbook can be developed by reference to the relevant sections in Chapters 5–7. They should include adequate arrangements for the management of any interfaces between contracts or work packages. It is especially vital to have procedures for:

- the transition of commissioning data record drawings and operating and maintenance manuals from one contract to the next

- confirmation that all relevant handover documentation and certification has been completed.

Engineering services commissioning

Engineering services commissioning is part of the construction stage. It is the main contractor's responsibility which is delegated to the services subcontractors. Action is taken in two stages: pre-contract and contract/post-contract.

Pre-contract

- Ensuring that the client recognises engineering services commissioning as a distinct phase of the construction process starting at the strategy stage.

- Ensuring that the consultants identify all services to be commissioned and defining the responsibility split for commissioning between designer, contractor, manufacturer and client.

- Identifying statutory and insurance approvals required and planning to meet requirements and obtain approvals.

- Co-ordinating the consultants' and client's involvement in commissioning to ensure conformity with the contract arrangements.

- Arranging single-point responsibility for control and the client's role in the commissioning of services.

- Ensuring contract documents make provision for services commissioning.

Contract and post-contract

- Ensuring relevant integration within construction schedules.

- Monitoring and reporting progress and arranging corrective action.

- Ensuring provision and proper maintenance of records, test results, certificates, checklists, software and drawings.

- Arranging for or advising on maintenance staff training, post-contract operation and specialist servicing contracts.

Examples of a checklist and documentation are given in Appendices C and D.

Completion and handover

The closely interlinked processes of completion and handover are very much a hands-on operation for the project manager and his team. This stage provides the widest and closest involvement with the client. Completion and handover require careful attention because they determine whether or not the client views the whole job as successful.

Completion

Handbook procedures may cover two sorts of agreement:

- Agreements for partial possession and phased (sectional) completion (if required):

  ○ access inspections, defects, continuation of other works and/or operation of any plant/services installation material, obstructions or restrictions.

  ○ certification on possession of each phase; responsibility for insurance.

- Agreements and procedures associated with practical completion:

  ○ user/tenant responsibility for whole of the insurance

  ○ provision within a specified time limit of complete sets of as-built and installed drawings, mechanical and engineering and other relevant installations/services data as well as all operating manuals and commissioning reports

  ○ storage of equipment/materials except those required for making good any defects

  ○ access for completion of minor construction works, rectification of defects, testing of services, verification of users' works and other welfare and general facilities.

Appendix 18 provides a typical checklist at the practical completion stage.

Handover

Procedures are needed for the following activities:

- To ensure that handover only takes place when all statutory inspections and approvals have been satisfactorily completed and subsequently to arrange that all outstanding works and defects are resolved before expiry of the defects liability period.

- To provide and agree a countdown schedule with the project team (examples of handover inspections and certificates checklists are given in Appendix E).

- To define responsibilities for all inspections and certificates.

- To monitor and control handover countdown against the schedule.

- To control pre-handover arrangements if the client has access to the building before handover.

- To deal with contractors who fail to execute outstanding works or correct defects including the possibility of implementing any contra-charging measures available under the contract. Agree and set up a procedure for contra-charging.

- To monitor and control any post-handover works which do not form part of the main contract.

- To monitor and control outstanding post-completion work and resolution of defects which form part of the main contract.

- To manage the end of the defects liability period and implement relevant procedures.

- To establish arrangements for the final account issuing the final certificate and carrying out the post-completion review/project evaluation report.

Client commissioning and occupation

Client commissioning

Client commissioning will involve the following handbook procedures:

- Arranging the appointment of the commissioning team in liaison with the client and establishing objectives (time, cost and specifications) and responsibilities at the feasibility and strategy stages.

- Preparation of a comprehensive commissioning and equipment schedule.

- Arranging access to the works for the commissioning team and client personnel during construction including observation of engineering services commissioning.

- Ensuring co-ordination and liaison with the construction processes and consultants.

- Preparing new work practice manuals and in close liaison with the client's/ user's facilities management team, arranging staff training and recruitment/ secondment of additional staff (e.g. aftercare engineer to support the client during the initial period of occupancy).

- Deciding the format of commissioning test and calibration records.

- Renting equipment to meet short-term demands.

- Deciding quality standards.

- Monitoring and controlling commissioning progress and reporting to the client.

- Reviewing post-contract the operation of the building at six, nine and 12 months: improvements, defects, corrections and related feedback.

Appendix G contains a relevant checklist.

Occupation

Occupation can be part of the overall project or a separate project on its own. A decision to this effect is made at the strategy stage with the client or user. The separate stages of occupation are set out below. Figures 7.1–7.4 illustrate these procedures graphically and Appendix G provides an example of an occupation implementation plan.

Structure for implementation
In order to achieve the necessary direction and consultations, individuals and groups are appointed, e.g.

- project executive (client/occupier/tenant)

- occupation co-ordinator (project manager)

- occupation steering group of a chair, co-ordinator and functional representatives concerned with the overall direction for:

  ○ construction schedule

  ○ technology

  ○ space planning

  ○ facilities for removal

  ○ user representation

  ○ costs and budget outline

- senior representative meeting of a chair (functional representative on steering group), co-ordinator and senior representatives of a majority of employees concerned with consultations on:

  ○ space planning

  ○ corporate communications

  ○ construction schedule problems

  ○ technology

- local representative groups chaired by manager/supervisor of their own group concerned with consultation at locations and/or departmental levels in order to ensure procedures for regular communications.

Scope and objectives (regularly reviewed)
- Identification of who is to move (project executive).

- Agreement on placement of people in new locations (steering group).

- Decision on organisation of move (steering group):

  ○ all at once

  ○ several moves

  ○ gradual flow.

- Reviewing time constraints (steering group):

  ○ construction

  ○ commercial

  ○ holidays.

- Identification of risk areas, e.g.

		○	construction delays and move flexibility
		○	organisational changes
		○	access problems
		○	information technology requirements
		○	furniture deliveries and refurbishment
		○	retrofit requirements.

Methodology ■ Listing special activities needed to complete the move, e.g.

  ○ additional building work

  ○ communications during move

  ○ provision of necessary services and move support

  ○ corporate communications

  ○ removal administration

  ○ furniture procurement

  ○ removal responsibility in each location/department

  ○ financial controls

  ○ access planning.

■ Preparation of a task list for each special activity, confirmation of the person responsible and setting the schedule of project meetings.

■ Production of outline and subsequently detailed schedule.

Organisation and control ■ Steering group establishes 'move group' to oversee the physical move.

■ Production of 'countdown' schedule (move group).

■ Identification of external resources needed (move group), e.g.

  ○ special management skills

  ○ one-off support tasks

  ○ duplication of functions during move.

■ Reporting to the client external support needs and costs (steering group).

■ Preparation of monitoring and regular review of actual budget (steering group), e.g.

  ○ dual occupancy

  ○ special facilities

  ○ additional engineering and technology needs

  ○ planning and co-ordinating process

  ○ inflation

  ○ external resources

  ○ non-recoverable VAT

  ○ contingencies.

PART 2 APPENDICES

APPENDIX A Typical meetings and their objectives

Steering group/team
- to consider project brief, design concepts, capital budget and programmes
- to approve changes to project brief
- to review project strategies and overall progress towards achieving client's goals
- to approve appointments for consultants and contractors.

Project team
- to agree cost plan and report on actual expenditure against agreed plan
- to review tender lists, tenders received and decide on awarding work
- to report on progress on design and construction programmes
- to review and make recommendations for proposed changes to design and costs, including client changes; to approve relevant modifications to project programmes.

Design team
- to review, report on and implement all matters related to design and cost
- to determine/review client decisions
- to prepare information/report/advice to project team on (1) appointment of sub/specialist contractors; (2) proposed design and/or cost changes
- to review receipt co-ordination and processing of subcontractors' design information
- to ensure overall co-ordination of design and design information.

Finance group/team
- to review, monitor and report financial, contractual and procurement aspects to appropriate parties
- to prepare a project cost plan for approval by the client
- to prepare and review regular cost reports and cash flows, including forecasts of additional expenditure
- to review taxation matters
- to monitor the preparation and issue of all tender and contract documentation
- to review cost implications of proposed client and design team changes.

Project team (programme/ progress meeting)
- to provide effective communication between teams responsible for the various phases of the project
- to monitor progress and report on developments, proposed changes and programme implications
- to review progress against programmes for each stage/section of the project/ works and identify any problems
- to review procurement status
- to review status of information for construction and contractors' subcontractors' requests for information.

Project team (site meeting)
Main contractor report tabled monthly to include details on:
- quality control

- progress
- welfare (health, safety, canteen, industrial relations)
- subcontractors
- design and procurement
- information required
- site security
- drawing registers
- Reports/reviews (including matters arising at previous meetings) from:
  - architect
  - building services
  - facilities management
  - information technology
  - quantity surveyor.
- Statutory undertakings and utilities:
  - telephones
  - gas
  - water
  - electricity
  - drainage.
- Approvals and consents:
  - planning
  - Building Regulations
  - local authority engineer
  - public health department
  - others.
- Information:
  - issued by design team (architect's instructions issued and architect's tender activity summary)
  - required from design team
  - required from contractor.

APPENDIX B Selection and appointment of contractors
B1 Pre-tender process

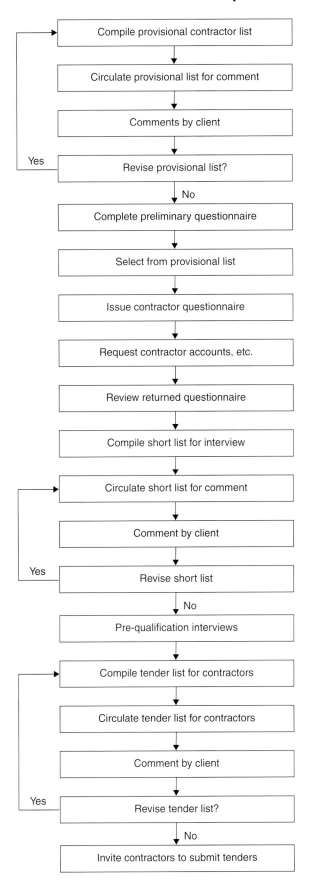

Checked by   Comments

APPENDIX B Selection and appointment of contractors
B2 Initial questionnaire

Project Management Ltd		Form Q1
Ref. number:		
Contract title:		
Item no.	**Question**	**Response**
1.0	Turnover of company?	
2.0	What is the value of contracts secured to date?	
3.0	What is the largest current contract?	
4.0	Is the contractor willing to submit a tender?	
5.0	Is the contractor willing to work with all team members?	
6.0	Is the contract period acceptable?	
7.0	If not how long to complete works?	
8.0	Is the anticipated tender period acceptable?	
9.0	If not how long to tender?	
10.0	What are the mobilisation periods of (a) completion of drawings? (b) fabrication? (c) start on site from order?	
11.0	Is the labour used direct self-employed or subcontract?	
12.0	What element of the contract will be sublet?	
Comments Signature and date _____		

APPENDIX B Selection and appointment of contractors
B3 Selection questionnaire

1 Name of company:
2 Address:
3 Telephone no.: Facsimile no.:
4 Nature of business:
5 Indicate whether (a) manufacturer (b) supplier (c) subcontractor (d) main contractor (e) design and build contractor (f) management contractor
6 Indicate whether (a) sole trader (b) partnership (c) private (d) public
7 Company registration number:
8 Year of registration:
9 Bank and branch:
10 VAT registration number:
11 Tax exemption certificate number: Date of expiry:
12 State annual turnover of current and previous four years:
13 State value of future secured work:
14 State maximum and minimum value of works undertaken:
15 Are you registered under BS 5750/ISO 9000?:
16 State previous projects undertaken with this company
17 Are you prepared to sign a design warranty?
18 Are you prepared to provide a performance bond?
19 Are you prepared to provide a parent company guarantee?
20 Do you operate a holiday with pay scheme?
21 State when stamps last purchased:
22 Do you contribute to the CITB?

APPENDIX B Selection and appointment of contractors
B3 Selection questionnaire (*cont'd*)

23 Do you have a safety policy?
24 Are you competent and willing to act as principal contractor under CDM?

25 Employer's liability insurance:
  Insurer:
  Policy no.:
  Expiry date:
  Limit of indemnity:

26 Third party insurance:
  Insurer:
  Policy no.:
  Expiry date:
  Limit of indemnity:

27 Which elements do you sublet?

28 List of projects of similar size and complexity:
 	
    Project 1.:
    Address:
    Architect:
    Contact: Telephone no.:
    Contractor:
    Contact: Telephone no.:
    Value:
    Year completed:
 	
    Project 2.:
    Address:
    Architect:
    Contact: Telephone no.:
    Contractor:
    Contact: Telephone no.:
    Value:
    Year completed:
 	
    Project 3.:
    Address:
    Architect:
    Contact: Telephone no.:
    Contractor:
    Contact: Telephone no.:
    Value:
    Year completed:

APPENDIX B Selection and appointment of contractors
B4 Pre-qualification interview agenda

Project Management Ltd	Form A1

Ref. number:

Contract title:

1.0 Introduction

1.1 Purpose of meeting

1.2 Introduction to those present

2.0 Description of overall project and schedule

2.1 General description of the project

2.2 Master schedule in summary

2.3 General description of contract

3.0 Explanation of contract terms and conditions

3.1 Outline and scope of contract

3.2 Responsibilities of the contractor

3.3 Outline of contract conditions including any significant amendments

3.4 Schedule

3.5 Specification

3.6 Drawings

3.7 Preliminaries

3.8 Budget prices

4.0 Project organisation

4.1 Site administration and project team

4.2 Setting out and dimensional control

4.3 Materials handling and control

4.4 Site establishment

4.5 Contractor supervision and on-site representative

4.6 Labour relations

4.7 Quality management

4.8 Health and safety plan

5.0 Tendering

5.1 Period of tendering

5.2 Mid-tender interview

5.3 Tender return date address and contact name

6.0 Actions required

6.1 Summary of actions and date deadlines

APPENDIX B Selection and appointment of contractors
B5 Tendering process checklist

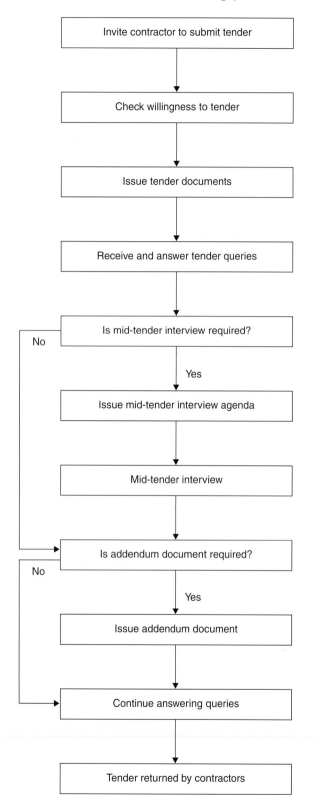

APPENDIX B Selection and appointment of contractors
B6 Tender document checklist

Project Management Ltd	Form C1

Ref. number:

Contract title:

☐ Invitation to tender

☐ Introduction and scope of contract

☐ Instructions to tenderers

☐ Form of tender

☐ General preliminaries

☐ Particular preliminaries

☐ Form of contract and amendments

☐ Contract schedule

☐ Method statement

☐ Quality management

☐ Project health and safety plan

☐ Project labour relations

☐ Specification

☐ List of drawings

☐ Bill of quantities or pricing schedule

☐ General summary

☐ Declaration of non-collusion

☐ Performance bond

☐ Warranty

☐ Soil report

☐ Contamination reports

Other documents (please list below)

☐

☐

☐

☐

☐

☐

☐

APPENDIX B Selection and appointment of contractors
B7 Mid-tender interview agenda

Project Management Ltd	Form A2

Ref. number:

Contract title:

1.0	**Introduction**	
1.1	Purpose of meeting	
1.2	Introduction of those present	

2.0	**Confirmation of addenda letters issued**	

3.0	**Responses to existing queries**	
3.1	Contractor	
3.2	Architect	
3.3	Civil and structural engineer	
3.4	Mechanical and electrical engineer	
3.5	Other consultants	
3.6	Quantity surveyors	
3.7	Project manager	

4.0	**Other additional information**	
4.1	Contractor	
4.2	Architect	
4.3	Civil and structural engineer	
4.4	Mechanical and electrical engineer	
4.5	Other consultants	
4.6	Quantity surveyors	
4.7	Project manager	

5.0	**Contractor's queries**	

6.0	**Confirmation of tender arrangements**	
6.1	Date	
6.2	Time	
6.3	Address	

7.0	**Any other business**	

APPENDIX B Selection and appointment of contractors
B8 Returned tender review process

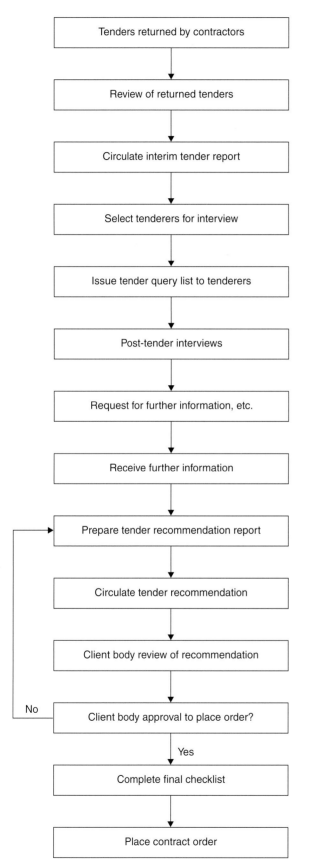

Part 2 Appendices

APPENDIX B Selection and appointment of contractors
B9 Returned tender bids record sheet

Project Management Ltd		Form R1	
Ref. number:			
Contract title:			
Allocated budget: £		**Programme period:**	
No.	**Contractor qualifications, etc.**	**Prog.**	**Bid sum**
1.			
2.			
3.			
4.			
5.			
6.			
7.			
8.			

Signed by the undersigned as a true record of duly and properly

Received tender bids for _____

On this day _____ (day) _____ (month) _____ in the year _____

Signed:

Company:

Signed:

Company:

Signed:

Company:

Signed:

Company:

APPENDIX B Selection and appointment of contractors
B10 Post-tender interview agenda

Project Management Ltd	Form A3

Ref. number:

Contract title:

1.0   Introduction

1.1   Introduction to those present

1.2   Purpose of meeting

2.0   Confirmation of contract scope and responsibilities

3.0   Detailed bid discussions

3.1   Contractual

3.2   Cost

3.3   Schedule

3.4   Method

3.5   Technical matters

3.6   Staffing labour and plant matters

3.7   Labour relations matters

3.8   Health and safety matters

3.9   Quality management

4.0   Contractor queries

5.0   Action and responses

5.1   Agreement of action items

5.2   Agreement of deadline dates for resolution of action items

APPENDIX B Selection and appointment of contractors
B11 Final tender evaluation report

Project Management Ltd	Form R2

Ref. number:

Contract title:

1.0	Summary of final tender bids following post-tender interviews
2.0	Cost appraisal
3.0	Schedule appraisal
4.0	Method statement appraisal
5.0	Technical appraisal
6.0	Contractual appraisal
7.0	Quality management appraisal
8.0	Health and safety appraisal
9.0	Labour relations appraisal
10.0	Recommendation to place contract

Appendices

1   Completed tender bid of recommended tenderer

2   Addenda and other information issues during tendering

3   Mid-tender interview meeting minutes

4   Query lists and responses

5   Post-tender meeting minutes

6   Any other letters, etc. since tender issue

7   Summary of contract buy and any other items to be bought

APPENDIX B Selection and appointment of contractors
B12 Approval to place contract order

Project Management Ltd	Form O1

Ref. number:

Contract title:

In accordance with clause _____ of the _____

_____ (Contract Form)

We _____

Do not have any objection to the placing of a contract with _____

for _____

All in accordance with the tender recommendation report submitted to us on the _____ (date).

In submitting the tender recommendation report the consultants are fully satisfied that the contractor has complied in full with the tender documents and is fully capable of carrying out the contract works.

Signed by: _____

Signed by: _____

Signed by: _____

APPENDIX B Selection and appointment of contractors
B13 Final general checklist

Project Management Ltd	Form C3

Ref. number:

Contract title:

Check once again that the following were carried out:

☐ Long list

☐ Telephone selection questionnaires

☐ Contractor selection questionnaires, company accounts, references and any reports of visits to offices, factories and contracts

☐ Short list

☐ Pre-qualification interview minutes

☐ Tender list

☐ Substitute tender list

☐ Tender documents and checklist

☐ Tender query lists addendum letters prior to mid-tender interviews

☐ Mid-tender interview minutes

☐ Tender query lists addendum letters, etc. post mid-tender interviews

☐ Returned tender summary form and returned tender documents

☐ Interim tender analysis and recommendations report

☐ Post-tender query lists to contractors

☐ Post-tender interview minutes

☐ Post-tender addendum letters, etc.

☐ Final tender analysis and recommendations report

☐ Contractor acceptability final check

☐ Approval to place contract order

APPENDIX C Engineering services commissioning checklist

Engineering services to be covered

- ■ Routinely:
  - ○ water supply and sanitation
  - ○ heating/cooling systems (boilers, calorifiers, chillers)
  - ○ ventilation systems
  - ○ air-conditioning
  - ○ electrical (generators, switchboards, others)
  - ○ mechanical (pumps, motors, others)
  - ○ fire detection and protection systems
  - ○ control systems (electrical, pneumatic, others)
  - ○ telephone/communications
- ■ specialist
  - ○ process plant for food, pharmaceutical, petrochemical or manufacturing activities
  - ○ security (CCTV, sensors, access control)
  - ○ facility management system
  - ○ acoustic and vibration scans
  - ○ lifts, escalators, others
  - ○ IT systems, e.g. IBM, DEC, ICL.

Contract documents

- ■ Responsibilities – client/contractor/manufacturer:
  - ○ bills of quantities/activity schedule items for commissioning activity with separate sums of clearly worded inclusion in M&E item descriptions
- ■ specification of commissioning:
  - ○ provision for providing that commissioning is performed – observation test results
  - ○ methods and procedures to be used, appropriate standards/codes of practice, e.g. CIBSE/IHVE/BSRIA/IEE/LPC/BS (see Appendix D).
- ■ provision for appropriate client access
- ■ client staff training
- ■ operating and maintenance manuals (as installed)
- ■ statutory approvals
- ■ record drawings and equipment software (as installed) and test certification
- ■ statutory approvals (lifts, fire protection, others)
- ■ insurance approvals.

Contractor's commissioning programme

- ■ Manufacturers' works testing
- ■ site tests prior to commissioning (component testing, e.g. a fan motor)

Part 2 Appendices

257

- pre-commissioning checks (full system, e.g. air-conditioning, by contractor before demonstration to client)

- set to work (system by system)

- commissioning checks (including balancing/regulation)

- demonstration to client (system basis)

- performance testing (including integration of systems)

- post-commissioning checks (including environmental fine-tuning during facility occupancy).

APPENDIX D Engineering services commissioning documents

CIBSE

Commissioning codes	A	Air distribution
	B	Boiler plant
	C	Automatic control
	R	Refrigerating systems
	W	Water distribution systems
	TM12	Emergency lighting

BSRIA

TM 1/88	Commissioning HVAC systems divisions of responsibilities
TN 1/90	European commissioning procedures
AG 1/91	The commissioning of VAV systems in buildings
AG 2/89	The commissioning of water systems in buildings
AG 3/89	The commissioning of air systems in buildings
AG 8/91	The commissioning and cleaning of water systems
AH 2/92	Pre-commissioning of BEMS – a code of practice
AH 3/93	Installation commissioning and maintenance of fire and security systems

HMSO

| HTM 17 | Health building engineering installations commissioning and associated activities (hospitals) |
| HTM 82 | Fire safety in health care premises fire alarms and detection systems (hospitals) |

Loss Prevention Council

| LPC | Rules for automatic sprinkler installations |

IEE

Wiring regulations

British Standards

An extensive list of BS publications exists for specialised systems and equipment, e.g. gas flues, steam and water boilers, oil and gas burning equipment, electrical equipment, earthing machinery, etc.

IT

Cabling installation and planning guide of relevant 'equipment' manufacturer/ supplier

APPENDIX E Handover checklists

Handover procedure

- architect's certificate
- Certificate of Practical Completion
- CDM health and safety file
- inspections and tests
- copies of certificates, approvals and licences
- release of retention monies
- final clean and removal of rubbish
- handover of spares
- meters read and fuel stocks noted.

Schedule

- remedial works
- Defects Liability Period and defect correction
- adjustment of building services
- client's fitting out.

Building owner's manual

- consultant's contributions
- format.

Operating and maintenance manuals, as-built drawings and C&T records

- servicing contracts established
- handover to facilities manager.

Letting or disposal

- schedule
- publicity
- strategy
- liaison
- documentation
- insurance.

Additional works

- contracts
- major service installations or adaptations
- fitting out
- shop fitting.

Final account, final inspection and Final Certificate arrangements

Liaison with tenants, purchaser or financier

Access by contractors
- remedial works
- additional contracts

Security
- key cabinet
- key schedule.

Publicity

Opening arrangements

Client's acceptance of building

Post-completion review/project close-out report

Inspection Certificates and Statutory Approvals

Fire officer inspections
- fire shutters
- fireman's lift
- smoke extract system/pressurisation
- foam inlet/dry riser
- fire dampers
- alarm systems
- alarm panels
- telephone link
- fire protection systems:
  - sprinklers
  - hose reels
  - hand appliances/blankets, etc.
- statutory signs.

Fire Certificate

Institution of Electrical Engineers' certificate

Water authority certificate of hardness of water

Insurer's inspections
- fire protection systems:
  - sprinkler
  - hose reels
  - hand appliances
- lifts/escalators
- mechanical services:
  - boilers

      ○   pressure vessels

      ○   electrical services

      ○   security installations

Officers of the court inspection (licensed premises)

Pest control specialists' inspection

Environmental health officer inspection

Building control officer inspection

Planning   ■   outline

               ■   detailed including satisfaction of conditions

               ■   listed building

Landlord's inspection

Health and safety officer's inspection

Crime prevention officer's inspection

APPENDIX F Client commissioning checklist

Brief	Ensure roles and responsibilities for commissioning team are developed and understood progressively from the feasibility and strategy stages.

Budget schedule
- based upon a clear understanding and agreement of the client's objectives.

Commissioning action checklist
- investigate and identify commissioning requirements
- management control document.

Appointments
- commissioning team
- operating and maintenance personnel
- aftercare engineer
- job descriptions, time-scales and outputs must be documented and agreed.

Client operating procedures
- work practice standards
- health and safety at work requirements.

Training of staff
- services
- security
- maintenance
- procedures
- equipment.

Client equipment (including equipment rented for commissioning)
- schedule
- selection
- approval
- delivery
- installation.

Building services and equipment
- define/check standards required in tender specification
  - testing
  - balancing
  - adjusting } detail format of records
  - fine tuning
- marking and labelling, including preparation of record drawings
  - handover of spares } must be compatible with any planned maintenance or equipment standardisation policies
  - handover of tools

Maintenance
- acceptance by client's maintenance section from the client's construction and commissioning team
- arrangements
- procedures
- contracts.

Security	■	alarm systems
	■	telephone link
	■	staff routes
	■	access (including card access)
	■	fire routes
	■	bank cash dispensers.
Communications	■	telephones
	■	radios
	■	paging
	■	public address systems
	■	easy-to-read plan of building
	■	data links.
Signs and graphics	■	code of practice for the industry
	■	statutory notices – H&S, fire, Factories Act, unions.
Initiation of operations	■	final cleaning
	■	maintenance procedures (including manufacturers' specialist maintenance)
	■	cleaning and refuse collection
	■	insurance required by date and extent of cover will vary with the form of contract
	■	access and security (including staff identity cards)
	■	safety
	■	meter readings or commencement of accounts for gas, water, electricity, telephone and fuel oil
	■	equipping
	■	staff 'decanting'
	■	publicity
	■	opening arrangements.
Review operation of facility	■	at +6, 9 and 12 months (including energy costs)
	■	improvements and system fine-tuning
	■	defects reporting, correction and verification procedures
	■	latent defects.
Feedback	■	channelled through aftercare engineer if appointed.

Glossary

Throughout this *Code* words in the masculine also mean the feminine and vice versa. Words in the plural include the singular, e.g. 'subcontractors' could mean just one subcontractor.

Aftercare engineer
The aftercare engineer provides a support service to the client/user during the initial 6–12 months of occupancy and is, therefore, most likely a member of the commissioning team.

Budget
Quantification of resources needed to achieve a task by a set time, within which the task owners are required to work. Note: a budget consists of a financial and/or quantitative statement, prepared and approved prior to a defined period, for the purpose of attaining a given objective for that period.

Business case
Information necessary to enable approval, authorisation and policy-making bodies to assess a project proposal and reach a reasoned decision.

Change order
An alternative name for variation order, it indicates a change to the project brief.

Change control
A process that ensures potential changes to the deliverables of a project or the sequence of work in a project, are recorded, evaluated, authorised and managed.

Client
Entity, individual or organisation commissioning and funding the project, directly or indirectly.

Client advisor
An independent construction professional engaged by the client to give advice in the early stages of a project, as advocated by the Latham Report.

Commissioning team
Client commissioning: predominantly the client's personnel assisted by the contractor and consultants. *Engineering services commissioning*: specialist contractors and equipment manufacturers monitored by the main contractor and consultants concerned.

Consultants
Advisors to the client and members of the project team. Also includes design team.

Contingency plan
Mitigation plan alternative course(s) of action devised to cope with project risks.

Contractor
Generally applied to: (a) the main contractor responsible for the total construction and completion process; or (b) two or more contractors responsible under separate contractual provisions for major or high technology parts of a very complex facility. (see **Subcontractor**).

Design audit
Carried out by members of an *independent* design team providing confirmation or otherwise that the project design meets, in the best possible way, the client's brief and objectives.

Design freeze
Completion and client's final approval of the design and associated processes, i.e. no further changes are contemplated or accepted within the budget approved in the project brief.

Design team   Architects, engineers and technology specialists responsible for the conceptual design aspects and their development into drawings, specifications and instructions required for construction of the facility and associated processes.

End user   Organisation or individual who occupies and operates the facility and may or may not be the client.

Facilities management   Planning, organisation and managing physical assets and their related support services in a cost-effective way to give the optimum return on investment in both financial and quality terms.

Facility   All types of constructions, e.g. buildings, shopping malls, terminals, hospitals, hotels, sporting/leisure centres, industrial/processing/chemical plants and installations and other infrastructure projects.

Feasibility stage   Initial project development and planning carried out by assessing the client's objectives and providing advice and expertise in order to help the client define more precisely what is needed and how it can be achieved.

Handbook   See **Project handbook**.

Life-cycle costing   Establishes the present value of the total cost of an asset over its operating life, using discounted cash flow techniques, for the purpose of comparison with alternatives available. This enables investment options to be more effectively evaluated for decision-making.

Master programme   This is the name given under some forms of contract to the baseline schedule, against which progress is expected to be monitored. It bears no relationship to the concept of the dynamic working schedule, used as a time model for the purposes of time management.

Occupation   Sometimes called *migration* or *decanting*. It is the actual process of physical movement (transfer) and placement of personnel (employees) into their new working environment of the facility.

Planning   The determination and communication of an intended course of action incorporating detailed method(s) showing time, place and resources required.

Planning gain   A condition attached to a planning approval which brings benefits to the community at a developer's expense.

Planning supervisor   A consultant or contractor appointed by a client under the CDM Regulations to carry out this role.

Principal contractor   The contractor appointed by a client under the CDM Regulations to carry out this role.

Programme management   A programme of works comprises a number of projects that are related because they contribute to a common outcome. Programme management provides co-ordinated governance to the realisation of benefits that result from projects; it is concerned with initiating projects, managing the interdependencies between projects, managing risk, and resolving conflicting priorities and resources across the projects.

Project   Unique process, consisting of a set of co-ordinated and controlled activities with start and finish dates, undertaken to achieve an objective conforming to specific requirements, including constraints of time, cost and resources.

Project brief   Statement that describes the purpose, cost, time and performance requirements/constraints for a project.

Project execution plan	A plan for carrying out a project, to meet specific objectives, that is prepared by or for the project manager. In some instances this is also known as the project management plan.
Project handbook	Guide to the project team members in the performance of their duties, identifying their responsibilities and detailing the various activities and procedures (often called the project bible). Also called project execution plan, project manual and project quality plan.
Project insurance	Project insurance is the descriptive title for a suite of insurances that are specifically designed to meet the needs of individual projects as opposed to relying on the individual insurance arrangements of the project team.
Project manager	Individual or body with authority, accountability and responsibility for managing a project to achieve specific objectives.
Project schedule	Time plan for a project or process. Note: on a construction project this is usually referred to as a 'project programme'. The construction industry tends to refer to programmes rather than schedules. Indeed the term schedule tends to mean a schedule of items in tabular form, e.g. door schedule, ironmongery schedule, etc.
Project sponsor	The project sponsor represents the client (which is usually the government) acting as a single focal point of contact with the project manager for the day-to-day management of the interests of the client organisation.
Project team	Client, project manager, design team, consultants, contractors and subcontractors.
Risk	Combination of the probability or frequency of occurrence of a defined threat or opportunity and the magnitude of the consequences of the occurrence.
Risk analysis	Systematic use of available information to determine how often specified events may occur and the magnitude of their likely consequences.
Risk factor	Associated with the anticipation and reduction of the effects of risk and problems by a proactive approach to project development and planning.
Risk management	Systematic application of policies, procedures, methods and practices to the tasks of identifying, analysing, evaluating, treating and monitoring risk.
Risk register	Formal record of identified risks.
Strategy stage	During this stage a sound basis is created for the client on which decisions can be made allowing the project to proceed to completion. It provides a framework for the effective execution of the project.
Subcontractor	An individual or company to whom the contractor sublets the whole or any part of the works. This covers such elements as design, specialist trades and labour-only supply.
Tenant	Facility user who is generally not the client or the developer.
User	The ultimate occupier of the facility.

Bibliography

The following is not intended to provide a comprehensive guide to the vast amount of literature available. Rather it is intended to support readers by directing them to supplementary titles which will allow construction project management and the intertwined processes to be evaluated and understood within its appropriate context.

A Guide to Managing Health and Safety in Construction (1995), Health and Safety Executive

A Guide to Project Team Partnering (2002), Construction Industry Council

A Guide to Quality-based Selection of Consultants: A Key to Design Quality, Construction Industry Council

Accelerating Change – Rethinking Construction (2002), Strategic Forum for Construction

ACE Client Guide 2000, Association of Consulting Engineers

Achieving Excellence Through Health and Safety, Office of the Government Commerce

Adding Value Through the Project Management of CDM (2000), Royal Institute of British Architects

APM Competence Framework (2008), Association for Project Management, High Wycombe

APM Introduction to Programme Management (2007), Association for Project Management – PMSI Group, High Wycombe

Appointment of Consultants and Contractors, Office of the Government Commerce

Benchmarking, Office of the Government Commerce

Bennett, J (1985), *Construction Project Management*. Butterworth, London

Bennett, J and Peace, S (2006), *Partnering in the Construction Industry – A Code of Practice for Strategic Collaborative Working*, CIOB/Butterworth Heinemann

Bennis, WG and Nanus, B (1985), *Leaders: The Strategies for Taking Charge*, Harper & Row, New York

Best Value in Construction (2002), Royal Institution of Chartered Surveyors

Briefing the Team (1996), Construction Industry Board

Building a Better Quality of Life, A Strategy for More Sustainable Construction (2000), Department of Environment, Transport and the Regions/Health and Safety Executive

Burke, R (2001), *Project Management Planning and Control Techniques*, 3rd edn

Client Guide to the Appointment of a Quantity Surveyor (1992), Royal Institution of Chartered Surveyors

Code of Estimating Practice, 5th edn (1983), The Chartered Institute of Building

Code of Practice for Project Management for Construction and Development, 3rd edn (2002), The Chartered Institute of Building

Code of Practice for Selection of Main Contractors (1997), Construction Industry Board

Code of Practice for Selection of Subcontractors (1997), Construction Industry Board

Constructing Success: Code of Practice for Clients of the Construction Industry (1997), Construction Industry Board

Constructing the Team, Sir Michael Latham (1994), Final report of the Government/ industry review of procurement and contractual arrangements in the UK construction industry (the Latham Report), HMSO

Construction (Design and Management) Regulations 1994, Health and Safety Executive

Construction (Design and Management) Regulations 1994, HMSO

Construction (Health, Safety and Welfare) Regulations 1996, Health and Safety Executive

Construction Best Practice Programme (CBPP) Fact Sheets

Construction Health and Safety Checklist (Construction Information Sheet No. 17), Health and Safety Executive

Construction Management Contract Agreement (Client/Construction Manager) (2002), Royal Institute of British Architects

Construction Management Contract Guide (2002), Royal Institute of British Architects

Construction Project Management Skills (2002), Construction Industry Council

Control of Risk – A Guide to the Systemic Management of Risk from Construction (SP 125) (1996), Construction Industry Research and Information Association

Cox, A and Ireland, P (2003), *Managing Construction Supply Chains*, Thomas Telford, London

Dallas, M (2006), *Value & Risk Management – A Guide to Best Practice*, CIOB/Blackwell, Oxford

Earned Value Management: APM Guidelines (2008), Association for Project Management – EVMSI Group, High Wycombe

Essential Requirements for Construction Procurement Guide, Office of the Government Commerce

Essentials of Project Management (2001), Royal Institute of British Architects

Fielder, FE (1967), *A Theory of Leadership Effectiveness*, McGraw-Hill, New York

Financial Aspects of Projects, Office of the Government Commerce

Goleman, D (2000), Leadership that gets results. *Harvard Business Review*, March pril

Good Design is Good Investment. Advice to Client, Selection of Consulting Engineer, and Fee Competition (1991), Association of Consulting Engineers

Gray, C (1998), *Value for Money*, Thomas Telford

Green, D (ed.) (2000), *Advancing Best Value in the Built Environment – A Guide to Best Practice*, Thomas Telford, London

Guide to Good Practice for the Management of Time in Complex Projects (forthcoming), Chartered Institute of Building

Guide to Project Management BS 6079 – 1 (2000), British Standards Institution

Hamilton, A (2001), *Managing Projects for Success*, Thomas Telford, London

Interfacing Risk and Earned Value Management (2008), Association for Project Management – RSI Group, High Wycombe

Kotter, J (1990), *A Force for Change: How Leadership Differs from Management*, Free Press, New York

Langford, D, Hancock, MR, Fellows, R and Gale, AW (1995), *Human Resources Management in Construction*, Longman, Harlow

Lock, D (2001), *Essentials of Project Management*, Gower Publishing

Management Development in the Construction Industry – Guidelines for the Construction Professionals, 2nd edn (2001), Thomas Telford, London

Managing Health and Safety in Construction. Construction (Design and Management) Regulations 1994. Approved Code of Practice and Guidance (2001), HSG224 HSE Books, Health and Safety Executive

Managing Project Change – A Best Practice Guide (C556) (2001), Construction Industry Research and Information Association

Manual of the BPF System for Building Design and Construction (1983), British Property Federation

Mintzberg, H (1998), Covert leadership: notes on managing professional. *Harvard Business Review* Nov–Dec, pp. 140–147

Models to Improve the Management of Projects (2007), Association for Project Management, High Wycombe

Modernising Construction: Report by the Comptroller and Auditor General (2001), HMSO

Modernising Procurement: Report by the Comptroller and Auditor General (1999), HMSO

Morris PWG, *The Management of Projects* (1998), Thomas Telford, London

Murdoch, I and Hughes, W (1992), *Construction Contracts: Law and Management*, E & FN Spon, London

Murray-Webster, R. and Simon, P (2007), *Starting Out in Project Management*, 2nd edn, APM, High Wycombe

Northhouse, P (1997), *Leadership – Theory and Practice*, Sage, California

Partnering in the Public Sector – A Toolkit for the Implementation of Post-award, Project Specific Partnering on Construction Projects (1997), European Construction Institute

Partnering in the Team (1997), Construction Industry Board

Planning: Delivering a Fundamental Change (2000), Department of Environment, Transport and the Regions

Potts, K (1995), *Major Construction Works: Contractual and Financial Management.* Longman, Harlow

Prioritising Project Risks (2008), Association for Project Management – RSI Group High Wycombe

Procurement Strategies, Office of the Government Commerce

Project Evaluation and Feedback, Office of the Government Commerce

Project Management (2000), Royal Institute of British Architects

Project Management Body of Knowledge, 5th edn (2006), Association for Project Management

Project Management in Building, 2nd edn (1988), The Chartered Institute of Building

Project Management Memorandum of Agreement and Conditions of Engagement, Project Management Panel, RICS Books

Project Management Planning and Control Techniques (2001), Royal Institute of British Architects

Project Management Skills in the Construction Industry (1996), Construction Industry Council

Project Risk Analysis and Management Guide, 2nd edn (2004) Association for Project Management – RSI Group, High Wycombe

Quality Assurance in the Building Process (1989), The Chartered Institute of Building

Rethinking Construction – Report of the Construction Task Force to the Deputy Prime Minister on the Scope for Improving the Quality and Efficiency of UK Construction (the Egan Report) (1998), Department of Environment, Transport and the Regions

Risk Analysis and Management for Projects (1998), Institution of Civil Engineers and Institute of Actuaries

Safety in Excavations (Construction Information Sheet No. 8), Health and Safety Executive

Selecting Consultants for the Team (1996), Construction Industry Board

Selecting Contractors by Value (SP 150) (1998), Construction Industry Research and Information Association

Shackleton, V (1995), Business Leadership, Routledge, London

Sustainability and the RICS Property Lifecycle (2009), RICS Books

Teamworking, Partnering and Incentives, Office of the Government Commerce

The CIC Consultants' Contract Conditions, Scope of Services and Scope of Services Handbook (2007), RIBA Publishing

The Procurement of Professional Services: Guidelines for the Application of Competitive Tendering (1993), Thomas Telford, London

The Procurement of Professional Services: Guidelines for the Value Assessment of Competitive Tenders (1994), Construction Industry Council

Thinking About Building? Independent Advice for Small and Occasional Clients, Confederation of Construction Clients

Thompson, P and Perry, JG (1992), *Engineering Construction Risks – A Guide to Project Risk Analysis and Risk Management*, Thomas Telford, London

Tichy, N and Devanna, M (1986), *The Transformational Leader*

Turner, JR (1999), *The Handbook of Project-based Management*, McGraw-Hill

Value by Competition (SP 117) (1994), Construction Industry Research and Information Association

Value for Money in Construction Procurement, Office of the Government Commerce

Value Management in Construction: A Client's Guide (SP 129) (1996), Construction Industry Research and Information Association

Walker, A (2002), *Project Management in Construction*, Blackwell, Oxford

Whole Life Costs, Office of the Government Commerce

Useful websites

Chartered Institute of Building – www.ciob.org.uk

Commission for Architecture and the Built Environment – www.cabe.org.uk

Confederation of Construction Clients – www.clientsuccess.org.uk

Construction Excellence – www.constructingexcellence.org.uk

Construction Industry Council – www.cic.org.uk

Construction Industry Research and Information Association – www.ciria.org

Construction Skills – www.cskills.org

Health and Safety Executive – www.hse.gov.uk

Housing Forum – www.thehousingforum.org.uk

Institution of Civil Engineers – www.ice.org.uk

Local Government Task Force – www.constructingexcellence.org.uk/sectorforums/lgtf/default.jsp

Movement For Innovation – www.constructingexcellence.org.uk/resources/az/view.jsp?print=true&id=290

National Audit Office – www.nao.gov.uk

Office of Government Commerce – www.ogc.gov.uk

Office of Public Sector Information/Her Majesty's Stationery Office – www.opsi.gov.uk

Royal Institute of British Architects - www.architecture.com

Royal Institution of Chartered Surveyors – www.rics.org.uk

Past Working Groups of *Code of Practice for Project Management*

Third Working Group for the Revision of the *Code of Practice for Project Management*

F A Hammond MSc Tech CEng MICE FCIOB MASCE FCMI — Chairman

Martyn Best BA Dip Arch RIBA — Royal Institute of British Architects

Alan Howlett CEng FIStructE MICE MIHT — Institution of Structural Engineers

Gavin Maxwell-Hart BSc CEng FICE FIHT MCIArb — Institution of Civil Engineers

Roger Waterhouse MSc FCIOB FRICS MSIB FAPM — Royal Institution of Chartered Surveyors and Association for Project Management

Richard Biggs MSc FCIOB MAPM MCMI — Association for Project Management

John Campbell — Royal Institute of British Architects

Mary Mitchell — Confederation of Construction Clients

Jonathan David BSc MSLL Services Engineers — Chartered Institution of Building

Neil Powling FRICS DipProjMan (RICS) — Royal Institution of Chartered Surveyors

Brian Teale CEng MICBSE DMS

David Trench CBE FAPM FCMI

Professor John Bennett FRICS DSc

Peter Taylor FRICS

Barry Jones FCIOB

Professor Graham Winch PhD MCIOB MAPM

Ian Guest BEng

Ian Caldwell BSc Barch RIBA ARIAS MIMgt

J C B Goring MSc BSc (Hons) MCIOB MAPM

Artin Hovsepian BSc (Hons) MCIOB MASI

Alan Beasley

David Turner

Colin Acus

Chris Williams DipLaw DipSury FCIOB MRICS FASI

Saleem Akram B Eng MSc (CM) PE FIE MASCE MAPM MACost E	– Secretary and Technical Editor of third edition
Arnab Mukherjee B Eng MSc (CM)	– Assistant Technical Editor
John Douglas	– Englemere Ltd

First and Second Working Groups of the *Code of Practice for Project Management*

F A Hammond MCs Tech CEng MICE FCIOB MASCE FIMgt	– Chairman
G S Ayres FRICS FCIArb FFB	– Royal Institution of Chartered Surveyors
R J Cecil DipArch RIBA FRSA	– Royal Institute of British Architects
D K Doran BSc Eng DIC FCGI CEng FICE FIStructE	– Institution of Structural Engineers
R Elliott CEng MICE	– Institution of Civil Engineers
D S Gillingham CEng FCIBSE	– Chartered Institution of Building Services Engineers
R J Biggs MSc FCIOB MIMgt	– Technical editor of second edition

MAPM Association for Project Management

J C B Goring BSc (Hons) MCIOB MAPM	
D P Horne FCIOB FFB FIMgt	
P K Smith FCIOB MAPM	
R A Waterhouse MSc FCIOB MIMgt MSIB MAPM	
S R Witkowski MSc (Eng)	– Technical editor of first edition
P B Cullimore FCIOB ARICS MASI MIMgt	– Secretary

For the second edition of the *Code* changes were made to the working group which included:

L J D Arnold FCIOB	
P Lord AA Dipl (Hons) RIBA PPCSD FIMgt (replacing R J Cecil, deceased)	– Royal Institute of British Architects
N P Powling Dip BE FRICS Dip Proj Man (RICS)	– Royal Institution of Chartered Surveyors
P L Watkins MCIOB MAPM	– Association for Project Managers

Index